Laurent Bortolotto

Direct hydroxylation of benzene to phenol in a microstructured Pd-based membrane reactor

Direct hydroxylation of benzene to phenol in a microstructured Pd-based membrane reactor

by
Laurent Bortolotto

Dissertation, Karlsruher Institut für Technologie
Fakultät für Chemieingenieurwesen und Verfahrenstechnik
Tag der mündlichen Prüfung: 06.05.2011
Referenten: Prof. Dr.-Ing. Roland Dittmeyer, Prof. Dr.-Ing. Georg Schaub

Impressum

Karlsruher Institut für Technologie (KIT)
KIT Scientific Publishing
Straße am Forum 2
D-76131 Karlsruhe
www.ksp.kit.edu

KIT – Universität des Landes Baden-Württemberg und nationales
Forschungszentrum in der Helmholtz-Gemeinschaft

KIT Scientific Publishing 2011
Print on Demand

ISBN 978-3-86644-695-3

Foreword

In early 2007, I bid farewell to the ever-changing and demanding world of the automotive industry in Stuttgart to join the Technical Chemistry workgroup, under the leadership of Prof. Dittmeyer, at DECHEMA in Frankfurt, as it provided me with the opportunity to return to research & development work and further develop my knowledge in the field of palladium membranes, which I had previously carried out research on at the DaimlerChrysler R&T of EADS in Friedrichshafen. Both promising and challenging, the aim of the project was to reproduce and study the much discussed gas phase direct hydroxylation of benzene to phenol in a Pd-membrane reactor, which was postulated as a new phenol route by the AIST researchers (Japan) in 2002. The project involved many activities from differing technical fields, such as reactor design (realization of a laboratory-scale planar reactor), materials engineering (membrane preparation on a commercial support), processes (dosage of gases through membranes into microstructured channels) and chemical reaction engineering (performance of a gas phase reaction over a metal catalyst) and in turn would allow me to expand my knowledge in these domains. The project DI696/6-1, which was financed by the German Research Foundation (DFG) and had a duration of 3 years, commenced in April 2007 based on the project proposal, this thesis represents the findings and conclusions determined over the course of the project. The project was supervised by Prof. Dittmeyer, at the DECHEMA in Frankfurt and later from the Karlsruhe Institute of Technology, whose extensive expertise on and in-depth research in the field of Pd-membranes for in-situ hydrogen extraction is internationally acknowledged.

Acknowledgements

A work of this magnitude would, of course, not be possible or at the very least be very difficult without the support of many persons both within and outside of work. I would particularly like to thank my fellow members of the Technical Chemistry workgroup and those members of the DECHEMA staff, who supported me during the period I worked on the DI696/6-1 project. I would like to expressly thank Prof. Dittmeyer for the invaluable support and advice, which he has provided me with over the course of this project as well as his commitment, which has, time and again, helped me gain new insights as well as push myself harder when it came to overcoming and dealing with the challenges and obstacles that arise as part of such a project. I would also like to mention Walter Jehle from DaimlerChrysler Research & Technology, who was the first person to introduce me to the field of Pd-membranes applied to onboard hydrogen extraction systems. I would also like to acknowledge the contribution of Mrs. U. Gerhards from IMVT (KIT, Karlsruhe) and Dr. M. Armbrüster (MPI, Dresden) with regard to the ESMA and XRD analysis of the membranes. Last but not least my family in France. Every achievement is for them and would be unattainable without them.

Abstract

The gas phase direct hydroxylation of benzene into phenol with hydrogen and oxygen, as initially described by Niwa in 2002, was realized in a newly developed double-membrane reactor. In the concept described by Niwa, a dense Pd membrane is used for the (safe) dosage of dissociated hydrogen species into the tubular reaction zone, where it reacts with gas phase oxygen to form surface species capable of directly converting benzene into phenol in a single-step process. In the flat reactor design used in this project, the second membrane, for the simultaneous dosage of oxygen, allows for a better control of the reaction atmosphere all along the distributed reaction channels. In this connection, a stable H_2/O_2 ratio favors the desired benzene hydroxylation conditions. Similar to other researchers in this field, the direct hydroxylation of benzene into phenol was also observed on a Pd surface (the process parameters T=150°C and H_2/O_2=1.4 over a Pd/PdCu membrane resulted in the highest selectivity and rate). The phenol rate with 1.3×10^{-8} mol/h is, however, far behind the CO_2 rate (range of 10^{-6} mol/h) and H_2O (range of 10^{-3} mol/h) rate, observed in previous research in this domain, which makes them the main products of the system. Furthermore, the C-based selectivity was limited to 9.6%.

The phenol performance of the reactor was improved by sputtering an active catalyst onto the surface of the PdCu membrane. The 3 systems studied in this connection were: PdGa, PdAu (both catalytic layers) and V_2O_5/PdAu (particles for a higher active surface), PdAu provided the highest C-based phenol selectivity (up to 67%) and rate (up to 7.2×10^{-7} mol/h). In terms of performance, the PdAu layer was the catalyst tested for phenol, which has the most positive result as part of this study. Most interestingly, the PdGa-modified surface was the catalytic system, which was the most successful in restricting the byproduct formation and this in spite of its reduced H_2-permeation properties and 10-fold lower phenol rates it achieved in comparison to the other two systems. We believe this is due to the isolated active sites on the surface of this intermetallic alloy as it is assumed that the adsorbed benzene molecules are only weakly bonded at these locations. In this sense, PdGa is a promising catalyst when it comes to lowering the formation rate of the reaction byproducts involved in the system. The influence of the main process parameters (temperature and partial pressures) was investigated with different catalytic surfaces and discussed.

Despite its technical achievements (proven feasibility of gas phase benzene direct hydroxylation and the advantages of the double-membrane dosage in comparison to tubular single-membrane systems), an inefficient use of hydrogen for obtaining phenol and the high gas phase stability of benzene (benzene conversion below 1%) unfortunately make the system impractical for an industrial application. Based on these observations, the mismatch between our results and those first mentioned by the reaction pioneers, which made industrial applications appear to be within grasp, must be stressed. This work, nonetheless, throws up some interesting aspects for other chemical engineering applications involving double-membrane dosage and the use of $Pd_{50}Ga_{50}$ at.% as a catalyst, which has the potential to reduce the influence of side-reactions in the presence of gas phase oxygen and hydrogen.

In addition to the experimental part, a system simulation was realized with Matlab 2007b. The evolution of the H_2/O_2 ratio due to the double-membrane dosage and the low benzene consumption were modeled in parallel to the product increase along the reactor axis. Measurements were carried out to correlate the simulation data with in-situ sample extraction by means of micro-capillaries implanted in the reactor sealing. A model based on adsorption and surface reaction mechanisms is proposed to describe the molecular mechanisms acting in the system and to enable a prediction of the product amounts in a larger process parameter range.

In light of the findings over the duration of the project, this work aims to contribute to a better understanding of the phenomena involved in the gas phase direct hydroxylation of a given aromatic compound, performed in a Pd-based membrane reactor. A topic, which has been much discussed since

2002. It also implements an innovative concept in the field of membrane reactors with the combined use of 2 gas distribution membranes in a single system.

Kurzfassung

Die direkte Hydroxylierung von Benzol zu Phenol in der Gasphase mit Wasserstoff und Sauerstoff, vorgestellt durch Niwa [24] im Jahre 2002, wurde in einem neuen Doppelmembranreaktor durchgeführt. In Niwas Konzept dient eine dichte Pd-Membran zur sicheren Dosierung von aktiviertem Wasserstoff in den Reaktionskanal. Dort reagiert der aktivierte Wasserstoff mit dem Gasphasen-Sauerstoff an der Pd-Membranoberfläche zu Hydroxyl-Radikalen. Diese können an der Membran adsorbierte Benzolmoleküle zu Phenol hydroxylieren.

In unserem planaren Design wird eine zweite Membran eingesetzt um auch die Zufuhr und Verteilung von Sauerstoff in den Reaktionskanälen besser kontrollieren zu können. Dadurch kann im gesamten Reaktionsbereich das optimale H_2/O_2-Verhältnis für die Benzolhydroxylierung ermittelt und eingestellt werden. Analog zu den Ergebnissen von Niwa konnte auch mit diesem Design die einstufige Hydroxylierung von Benzol zu Phenol erreicht werden. Die Variation der Reaktionsparameter für ein System mit Pd/PdCu-Membranen ergab ein Optimum hinsichtlich der Phenolrate und Selektivität bei 150°C und einem H_2/O_2-Verhältnis von 1.4. Mit einer Phenolrate von lediglich 1.3×10^{-8} mol/h ist die Phenolleistung jedoch weit unter der CO_2- (ca. 10^{-6} mol/h) und insbesondere der H_2O-Rate (ca. 10^{-3} mol/h). Wasser und CO_2 sind die Hauptprodukte des Systems. Hinzu kommt, dass die C-basierte Phenolselektivität unter diesen Bedingungen nur einen Wert von 9.6% erreicht.

Durch das Sputtern von aktiven Elementen auf die PdCu Membran konnte die Phenolleistung des Systems verbessert werden. Es wurden drei Systeme untersucht: PdGa- und PdAu-Schichten sowie V_2O_5/PdAu-Partikel zur Vergrößerung der katalytisch aktiven Oberfläche. Die PdAu-Schicht hat die beste Phenolrate (max. 7.2×10^{-7} mol/h) und -selektivität (max. 67%) erreicht. Trotzt der 10-fach niedrigeren Phenolrate gegenüber PdAu und der niedrigen H_2-Permeanz konnte die katalytische PdGa-Schicht die Bildungsrate der

Nebenprodukte am besten unterdrücken, da die Anwesenheit von isolierten aktiven Zentren auf der Oberfläche die Adsorptionseigenschaften von Benzol beeinflusst. PdGa gilt als vielversprechender Katalysator hinsichtlich der Einschränkung von Nebenreaktionen. Der Einfluss der Hauptprozessparameter, Temperatur und Partialdrücke, wurde an den genannten katalytischen Oberflächen ebenfalls untersucht.

Trotz der technischen Vorteile des Systems; Reproduzierbarkeit der direkten Hydroxylierung in der Gasphase und positiver Einfluss der 2. Membran als Weiterentwicklung des Monomembran-Reaktorsystems, ist dieses Verfahren für eine industrielle Anwendung ungünstig, insbesondere wegen seiner sehr niedrigen Phenolausbeute hinsichtlich des weit überstöchiometrischen Wasserstoffverbrauchs. Reaktionshemmend wirkt sich vor allem die hohe Stabilität von Benzol in der Gasphase aus. Diese Untersuchungen hinsichtlich des Anwendungspotentials des Verfahrens weisen große Unterschiede zu denen 2002 von Niwa publizierten Ergebnissen auf. Dennoch eröffnet diese Studie interessante Perspektiven für die Anwendung des Doppelmembran-Konzepts und des $Pd_{50}Ga_{50}$-Katalysators zur verbesserten Reaktionskontrolle.

Neben dem experimentellen Anteil der Arbeit wurde das System mit Hilfe von Matlab 2007b simuliert, um die Edukt- und Produktkonzentrationsprofile entlang des Reaktors abschätzen zu können. Mit Matlab wurde sowohl das Produktbildungsprofil und die Stabilität des H_2/O_2-Verhältnisses entlang der Reaktorachse als auch der geringe Nutzungsgrad von Benzol in der Gasphase simuliert. Zur Verifizierung der Simulation wurden Vergleichsmessungen anhand von Gasprobenextraktion mit Kapillaren durchgeführt. Ein LANGMUIR-HINSHELWOOD-basiertes Reaktionsmodel wurde vorgeschlagen, um die involvierten Oberflächenmechanismen beschreiben und die Produktzusammensetzung für andere Prozessparameter abschätzen zu können. Die vorliegende Arbeit liefert ein besseres Verständnis der Reaktionsmechanismen und der resultierenden Produktbildung, die während der direkten Hydroxylierung von Benzol in einem Pd-Membranreaktor auftreten. Die Einführung des 2-Membranenkonzepts liefert hier eine bessere

Kontrolle der Reaktionsbedingungen verglichen mit dem ursprünglichen Konzept von Niwa.

Table of contents

Table of contents

XII

1. Project motivation

Some commodity products, which are produced on a daily basis by the chemical industry, can be found in not just one but several industrial sectors, where they are used as base materials and transformed into more valuable products. This is the case for the aromatic chemical phenol (formula C_6H_5OH), which is a white powder in its physical state at room temperature and characterized by its benzene ring. Phenol is commonly employed in the pharmaceutical, automotive and construction industries where it is processed to obtain medicines, disinfectants, resins for light composite materials and plywood products, amongst other things, respectively. The worldwide phenol production was estimated to be 6.6 megatons in 2000 [24].

The common industrial process for phenol, known as the Hock process or so-called cumene process, involves 3 manufacturing steps. Two products are obtained from this process, namely phenol and acetone as depicted in Fig.1-1:

- Reaction between benzene and propylene to form cumene (step 1). This step is realized at high pressure and at a temperature of 250°C, most of the time in the presence of phosphoric acid.
- Oxidation of cumene into cumene hydroperoxide in the presence of a radical initiator (step 2).
- Decomposition of cumene hydroperoxide into phenol and acetone in an acidic medium (step 3).

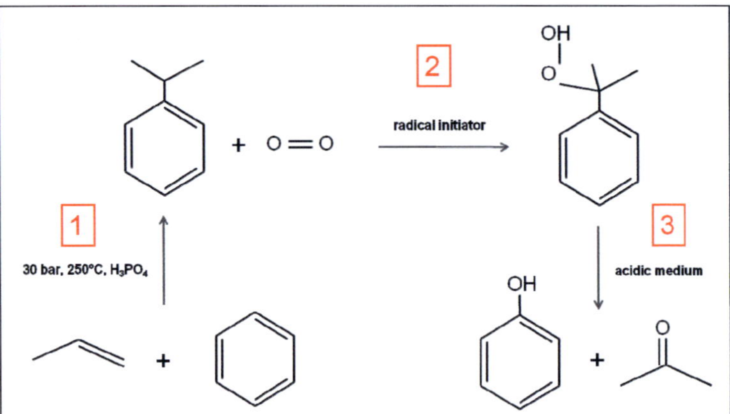

Fig. 1-1: Diagram of the different steps of the Hock process leading to the production of phenol and acetone from benzene and propylene

An additional distillation is then required to recover the phenol. Due to the multiple steps, this process achieves low phenol yields of around 5%, in addition to being complex and energy consuming. Moreover, whether the process is economic strongly depends on the acetone demand in the chemical market. These drawbacks, combined with high energy consumption, are the driving force for active research on new phenol production routes in different countries with the objective of developing a direct, efficient and cost-competitive process. These 3 requirements continue to remain unfulfilled as cost or safety issues are a barrier to any industrial prototyping of technical alternatives.

Research projects, whose objectives were to develop alternatives to the industrial Hock process and which were investigated by several research institutes and universities to this end, will be discussed in detail in the next chapter. The different methods examined as part of these projects have several advantages, for example, high reaction efficiency due to a satisfactory oxidant/catalyst pairing or process simplification in comparison to the cumene route for obtaining phenol. Additional systems have to be proposed and studied, however, as none of these methods have become a serious alternative for phenol production due to economic viability reasons as well as safety in long-term operation. This observation is the starting point and represents the motivation for this work.

One of the approaches suggested to simplify the phenol process is based on the use of a palladium membrane. Pd acts as a catalyst, forming reactive surface species from the hydrogen permeating through the membrane and the oxygen from the gas phase; these, in turn, react with benzene to form phenol. At the same time, the membrane keeps the hydrogen and oxygen separate thus improving the safety of the process. The Technical Chemistry working group of DECHEMA e.V. in Frankfurt/Main proposes and develops innovative uses of Pd-membrane technologies in process engineering and chemical reaction engineering as a matter of course under the direction of Prof. R. Dittmeyer. The following objectives were set, against this background, at the outset of the project. The direct phenol route – presented further on in this document – using benzene and gas phase oxygen and hydrogen in a Pd-membrane reactor will be investigated in detail with a focus on the following two aspects:

- ***The use of 2 membranes*** in the reactor for the combined dosage of both hydrogen and oxygen in the reaction area. This technical implementation, compared to a single Pd-membrane reactor, aims to provide a higher surface area in the reactor, which can be utilized for the hydroxylation of benzene into phenol. A competition between the hydroxylation and side-reactions has been pointed out by the research group that first introduced the concept of benzene direct hydroxylation in a Pd-membrane reactor; this is what motivated this concept improvement. The use of selective metal membranes is preferred as they would allow the use of mixed gases (for instance air or reformate gas) and would thus be more favorable in terms of realizing possible future applications at lower costs.

- ***The catalytic modification of the Pd-surface*** by sputtering active elements, where Pd is present in alloyed form, in order to increase the phenol selectivity in the reactor. The choice of the new catalytic systems for surface modification is based on the benefits observed by several researchers where they used said for liquid-phase or gas-phase reactions involving hydrogen.

They will be described and discussed in the respective chapters, with the emphasis on the benefits and limitations observed when applying them to the gas-phase hydroxylation of benzene into phenol in a Pd-membrane reactor. A discussion based on the experimental results, observations and system analysis shall also be presented.

2. Literature analysis

Due to the complexity of the industrial production process for obtaining phenol from benzene, many researchers have worked on alternatives solutions over the past decades, with the aim of realizing a direct synthesis of phenol from benzene, using a wide range oxidants and catalysts for the reaction. The following literature analysis aims to summarize these developments, from earlier examples of phenol direct synthesis using conventional process technology, with its advantages and drawbacks, to more recent works on phenol production in a Pd membrane reactor, with its potential and challenges.

2.1. Production of phenol from direct conversion of benzene

A key step to achieve the direct synthesis of phenol from benzene is to generate active oxygen species capable of breaking a C-H bond for hydroxylation. Different forms of oxygen species feature this required activity, such as the negatively charged O^-, O_2^-, O^{2-}, which results in an electrophilic reaction mechanism, or neutral oxygen species like the HO* or HOO* radicals. Many publications contain reports on efficient oxidants and catalysts used in direct conversion reactions for producing phenol from benzene. They can be divided into three categories.

2.2. Use of N_2O as an oxidant

Various sources have reported the use of nitrous oxide to supply the active oxygen [1-8]. Iwamoto et al. [1] appear to be the pioneers in the early 1980s, where they employed N_2O over a V_2O_5-SiO_2 catalyst and reported a benzene conversion of 11% and a phenol selectivity of 45%. Since then, the use of zeolites, like H-ZSM-5, as an efficient catalyst for nitrous oxide activation has been an important research field [2]. The development of zeolites led to a performance improvement as shown in later works, with a benzene conversion of 27% and a phenol selectivity of 99% as reported by

Ribera et al. using a Fe-ZSM-5 catalyst [8]. The properties of the Fe-zeolite catalyst on N_2O have been widely investigated since then, with more than 100 articles published on this topic in this decade alone. The good performances shown through the use of nitrous oxide for selective oxidation of benzene to phenol did not lead, however, to an industrial application due to the high N_2O supply cost.

2.3. Use of H_2O_2 as an oxidant

Studies have been carried out on the conversion of benzene to phenol in the liquid phase using hydrogen peroxide as oxidant [9-15], for example over vanadium oxide or titanium-silicalite-based catalysts [11, 12]. Phenol yields between 6 and 14% were reported in the liquid phase with a reaction temperature ranging from 30 to 60°C [10-12]. In later studies, Tang and Zhang employed H_2O_2 over a vanadium-substituted heteropolymolybdic acid catalyst at 70°C and achieved a phenol selectivity of 93% and a phenol yield of 10.1% [15]. While the use of hydrogen peroxide as an oxidant for benzene hydroxylation presents interesting results, it has the same limitation as N_2O for a large-scale production of phenol, with its high supply cost making the reaction economically not viable at this stage.

2.4. Use of H_2 and O_2 in the gas phase over a bi-functional catalyst

Due to the cost issues raised by the use of H_2O_2 and N_2O for phenol production, cheaper alternatives have to be found. The direct hydroxylation of benzene to phenol in the gas phase in the presence of both hydrogen and oxygen, over a bi-functional catalyst has been proposed and studied in several locations around the world.

Jintoku et al. used hydrogen and oxygen in the late 1980s over Si-based catalysts to produce phenol from benzene. Bi-functional catalysts like Pt/SiO_2, $Cu-Pt/SiO_2$ or $Pt-V_2O_5/SiO_2$ have been widely used [16-19] to produce phenol in the gas phase. Kunai et al. reported that phenol was obtained by the hydroxylation of benzene with O_2 and H_2 with almost 100% selectivity on a Cu-

Pd/SiO$_2$ catalyst at 60°C and atmospheric pressure [17]. However, the phenol yields obtained (ca. 15 µmol/h) appeared to be inadequate for a practical application.

The need for a greater efficiency of the bi-functional catalysts for this gas phase reaction has been proven. Kitano et al. [20] tested a Pd-Cu$^{I/II}$ catalyst deposited on a silica-gel support for benzene conversion. They observed in this case that if one of the two metal species were missing from the catalyst then no catalytic activity appeared. In contrast to the same reaction in the liquid phase, the use of a Pd-single component catalyst is thus inefficient in the gas phase reaction. They also observed that the proportion of the benzene hydrogenation by-products was minimized in the gaseous reaction, whereas they amount to an appreciable proportion of the product selectivity in the liquid phase reaction.

Hwang et al. [21] studied this gas phase reaction over a bi-functional catalyst system combining Pd-zeolites with redox zeolites such as TS-1, Ti-MCM-41, V-MCM, Fe-Y or Fe-phtalocyanine/Y. All these systems exhibited a high activity for the hydroxylation of benzene to phenol using H$_2$ and O$_2$ under mild conditions, with the in-situ formation of H$_2$O$_2$. The same year, Miyake et al. investigated the influence of the addition of transition metal salts on the performance of a Pt/SiO$_2$ based catalyst on the direct hydroxylation of benzene with H$_2$ and O$_2$, and found the addition of vanadium compounds resulted in the most active systems. It is suggested that these systems improve the in-situ formation of H$_2$O$_2$, which produces hydroxyl radicals via a Fenton-type mechanism [22]. Laufer et al. have investigated the use of Pd and Pt containing heterogeneous acid catalysts such as zeolites or Nafion/silica composites to hydroxylate benzene in the presence of H$_2$ and O$_2$ [23]. They obtained their best results over a strong acidic Nafion silica composite carrier with 0.5 wt % Pd and 0.5 wt % Pt at temperatures between 30 and 40°C.

Due to the low phenol yields obtained and taking into account the reaction restriction related to the explosion limit of the oxygen/hydrogen mixture used in the above-mentioned systems, safer and innovative alternatives have been proposed, such as the use of a Pd-membrane reactor to perform benzene hydroxylation. In this case, the Pd-membrane acts as a catalyst for the reaction and also as a diaphragm to prevent detonation due to the direct contact of hydrogen with oxygen.

2.5. Direct hydroxylation of benzene to phenol in a Pd-membrane reactor

The concept of the direct hydroxylation of benzene to phenol in a Pd-membrane reactor was first presented by Niwa et al. at the AIST in Japan in 2002 [24]. A system using a palladium membrane as catalyst for the oxidation of benzene to phenol was proposed in this case. In this concept, hydrogen molecules dissociate to atoms on the surface of the catalytic palladium membrane, dissolve into the membrane, diffuse through it and emerge on the reaction side in atomic form. Active hydrogen atoms should easily react with oxygen molecules to form active species like OH-radicals. It is assumed that the benzene is attacked by these radicals and readily hydroxylated to phenol. Fig. 2.5-1 illustrates the concept of the reductive oxidation of benzene to phenol using a catalytic Pd-membrane.

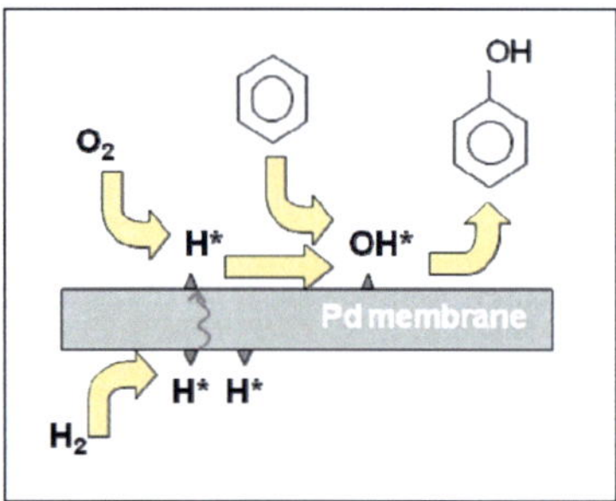

Fig. 2.5-1: Principle of benzene direct hydroxylation on a palladium membrane as presented by Niwa et al. [24]

This concept was proven by practical experiments. Five studies based on this catalytic system have been published so far by Japanese researchers, four on the direct hydroxylation of benzene to phenol in a Pd-membrane reactor [24, 25, 27, 28] and one on the direct hydroxylation of methyl benzoate to methyl salicylate in a Pd-membrane reactor [26]. All the experiments were carried out in tubular membrane reactors. Very encouraging results were initially reported in this system with a phenol yield exceeding 20% and a phenol selectivity above 77% attained at 200°C [25]. In this paper, phenol was accompanied mainly by cyclohexane and cyclohexanone, produced by phenol and benzene hydrogenation, respectively, but no information about the presence of total oxidation products, like CO_2, is provided. These results generated a strong interest as the reaction performance reported was very promising, even if they presented a low phenol production rate and produced a large amount of water, 500 to 1000 times the amount of phenol, while showing a low efficiency of hydrogen utilization.

The next publications reported, however, other results where the presence of by-products like CO_2 was not only noticed but dominant in the system. In 2004, Sato and Niwa reported on the direct hydroxylation of methyl benzoate to methyl salicylate using a Pd-membrane reactor, which was carried out in a similar manner [26]. The influence of the H_2/O_2 concentration ratio in the reactor on the direct hydroxylation of methyl benzoate to methyl salicylate was investigated. A large number of by-products were detected together with methyl salicylate. These included $CO_{2,}$, which was dominant in a large part of the reactor; hydrogenation products were also present close to the reactor end, restricting the yield of methyl salicylate to 4.7% at 150°C. It was speculated for the first time that hydroxylation only occurs in a limited area of the reactor where the atmosphere composition is adequate for the reaction [26]. The oxygen and hydrogen concentration profiles vary in opposite ways along the reactor length making the entrance part and end part of the reactor particularly suitable for oxidation and hydrogenation reactions, respectively. A year later, the same authors presented additional results on the direct hydroxylation of benzene to phenol in a Pd-membrane reactor, with a 1 μm Pd layer, carried out at temperatures between 150°C and 200°C [27]. The formation of a small amount of CO_2, not quantified, is mentioned this time. The phenol yield obtained was over 15% at 150°C with a phenol selectivity of 95%.

An increase in the reaction temperature led to an increase in benzene conversion but was also responsible for a decrease in phenol selectivity [27]. Finally, Sato et al., investigated the effects of membrane pretreatments on direct hydroxylation of benzene to phenol, with oxygen and hydrogen at 200°C [28]. The membrane pretreatment did not affect the hydroxylation activity, which is influenced by the concentration of H_2 and O_2; however the results reported show a lower reaction performance. After H_2 pretreatment, a benzene conversion of 10% in steady state and a phenol and quantified CO_2 yield of both 4% were obtained together with a yield of 2% of hydrogenation products. After O_2 pretreatment, CO_2 and byproducts were detected just after the reaction started whereas phenol did not appear before a reaction time of 300 min [28]. A favorable phenol production at these reaction conditions was no longer reported by this group due to the extent of the side-reactions in the system.

This tendency was confirmed in a joint study carried out on a similar system in the Netherlands in 2004 [29]. The concept of a Pd-based catalytic membrane reactor was also applied to the direct oxidation of benzene to phenol at 150°C. The formation of phenol via the in-situ H_2O_2 principle was proven, but experiments also showed evidence of the significant formation of by-products, namely the total oxidation of hydrogen to water and total oxidation of benzene to CO_2. The benzene conversion obtained was 3.8% with a phenol selectivity of only 4.1% [29]. The CO_2 selectivity was 95.9%. Vulpescu et al. evaluated the Pd-based catalytic membrane reactor system and came to the conclusion that it is not feasible for commercial application due to technical and economic reasons [29]. Another group from the Nanjing University of Technology (China) attempted to reproduce the gas phase hydroxylation of benzene to phenol in a Pd-membrane reactor using a thin Pd film produced by electroless plating [63]. Their experimental results showed that the dense Pd membrane was almost inert for the hydroxylation and oxidation of benzene (phenol yield below 0.13% where it was detected). They concluded that the concept for the one-step hydroxylation of benzene to phenol, with oxygen and hydrogen, was inefficient.

The contradictory results reported in Japan by the AIST compared to those in the Netherlands (DSM/ECN) and China (NUT), in addition to the innovative concept combining membrane technology and chemical reaction engineering, were another reason for being particularly interested in working on this reaction. This work represents a fourth detailed opinion on the potential of benzene hydroxylation in the presence of hydrogen and oxygen (in-situ H_2O_2) in a Pd-based membrane reactor. A new reactor concept for optimizing the safety and the concentration profiles in the reaction area is also proposed.

2.6. Metallic membranes

2.6.1. Hydrogen selective membranes based on Pd

The outstanding permeability properties of palladium vis-à-vis hydrogen are very well known and many articles have been published on this topic. Pd-based membranes are, in particular, widely used for gas separation, hydrogen removal [31-39] or hydrogen dosage, depending on the involved application. Pd can be alloyed with several metals, as presented by Knapton [30], to improve its permeability vis-à-vis hydrogen; this makes it more resistant to sulfur and prevents hydrogen embrittlement at low temperatures. In this work, a Pd-based membrane was used to dose hydrogen in a micro-channel reactor at a relatively low temperature (below 250°C). This application makes it necessary to alloy Pd with another metal in order to avoid hydrogen embrittlement below 295°C. An increased amount of hydrogen-rich β-phase, in fact, forms in the α-matrix (see Fig. 2.6-1) below 300°C as hydrogen permeates through the bulk of Pd. The higher hydrogen content of the β-phase compared to the α-phase leads to a ca. 5% lattice expansion, creating internal stresses in pure Pd films during hydrogen permeation, which may lead to crack formation or delamination of the Pd film from the support.

Based on Fig. 2.6-1, which shows the phase equilibrium of the Pd-H system, one can describe the phase formation in the palladium hydride depending on the process conditions (hydrogen partial pressure and temperature). The palladium retentate side of the membrane is subjected to the highest H_2 partial pressure as opposed to the permeate side (pressure gradient in the layer: for instance 2 bar as opposed to 0.024 bar). The temperature is constant during permeation. Only the α-phase is formed for any temperature above 295°C as hydrogen permeates through palladium, independently of the applied partial pressure. The H/Pd ratio increases during permeation but it will never reach the β-domain above the mentioned temperature. The α-phase is permeable to hydrogen, which diffuses through the layer without being "captured" in the Pd lattice to form hydrides (β-phase). For any temperature below 295°C (i.e. 250°C), the hydride β-phase will start to grow in the α-matrix at a certain stage of the diffusion upon reaching the α+β domain. This domain will be entered at a later stage (higher H/Pd ratio) in the case of a higher hydrogen partial pressure. The β-phase is saturated in hydrogen around the $PdH_{0.7}$ at.% composition whereas the α- phase is nearly hydrogen free. Knowing the used Pd volume and the amount of hydrogen dosed through the membrane for a given permeation time, one can assess the H/Pd molar ratio at the employed temperature (if below the critical temperature of 295°C). Where the α+β domain is reached, there is a risk of embrittlement of and damage to the Pd film.

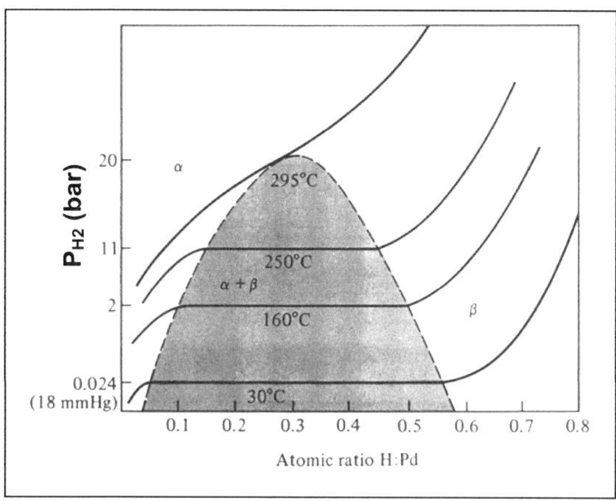

Fig. 2.6-1: Pd-H phase diagram with varying P_{H2} or varying temperatures (standard conditions apply to P_{H2} and T when they are the stable parameter)

Several alloying elements like Cu, Ag or Au prevent Pd against hydrogen embrittlement and even improve the transport properties of hydrogen through the Pd alloy film [30]. PdAg [40-42], PdCu [43-45] and PdAu [35] alloy layers are widely used and can be produced by several methods on commercial supports (electroless plating or sputtering). They are also commercially available as a finished product. As "bulk" material for the membrane, the PdAu alloy was excluded from the selection for cost reasons. Its use as a material for the dosage of hydrogen is not required due to the following two alternative alloys, namely $Pd_{77}Ag_{23}$ wt.% and $Pd_{60}Cu_{40}$ wt.%, which are more economic and have improved permeation properties in comparison to Pd. PdAu will, however, be subsequently considered as a catalyst [50; 51]. The selected "low temperature" Pd alloy is $Pd_{60}Cu_{40}$ wt.%. In spite of the sensitivity of its H_2 transport properties, depending on the Cu content (presence of both bcc and fcc phases in a narrow range of composition close to 60 wt.% Pd), it has a high permeability (1.52 times that of pure Pd) and a lower cost in comparison to $Pd_{77}Ag_{23}$ wt.%. Moreover, the positive effect of Cu in benzene hydroxylation has already been observed [17].

As shown in the Pd-Cu phase diagram of Fig. 2.6-2, a face centered cubic (fcc) solid solution of Pd and Cu is present in the whole composition range above 600°C. Below this temperature, a body centered cubic phase (bcc) exists between approx. 50 and 60 wt.% Pd. These composition values are the limits of the high H_2 permeable bcc domain before phase transition to fcc for an increasing Pd amount. As maximum permeability is obtained at 60 wt.% Pd, the composition has to be accurately produced in order to avoid the lattice transition in the narrow range shown in Fig. 2.6-2.

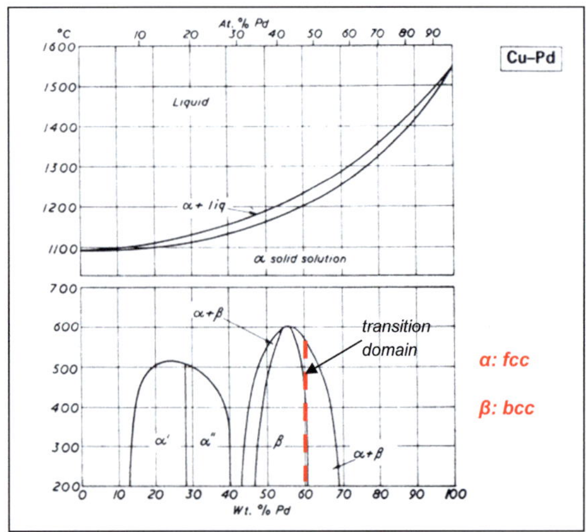

Fig. 2.6-2: Pd-Cu phase diagram

According to published data, supported $Pd_{66}Cu_{34}$ wt.% membranes (4 µm thick) are able to deliver a hydrogen flux of 0.19 mol/m^2/s at 510°C for a net hydrogen partial pressure difference of 0.35 bar between the retentate and permeate sides of the membrane [44]. Ignoring as a first approximation the influence of the composition difference of the alloy in comparison to a $Pd_{60}Cu_{40}$ wt.% film of the same thickness, one would expect a H_2 flux ranging from 3.0x10^{-2} mol/m^2/s to 4.2x10^{-2} mol/m^2/s through the membrane between 150°C and 250°C, respectively, for the same driving force. For an increased driving force of 1 bar, the permeating flux at 150°C would be around 1.2×10^{-1} $mol/m^2/s$.

2.6.2. Oxygen selective membranes

Silver

The use of a second membrane for the dosage of oxygen in the reactor in order to perform the direct hydroxylation of benzene to phenol with a better control of the reaction atmosphere led to the search for materials, which are permeable to this gas. In relation to possible metals for the hydrogen selective membrane, the element the most well-known for its oxygen transport properties was silver. Silver can be deposited as a layer on a substrate via electroless plating or purchased as a thin foil. The permeation properties of silver in relation to oxygen were examined in greater detail through the work of Cole on this system in the 1960s [47], followed by Gryaznov in the 1970s [46]. Saracco and Specchia also published comparative results on different types of oxygen selective membranes [60]. More recently, a study collecting and summarizing the different works on Ag-selective membranes was realized at the University of Utrecht [48]. Gryaznov reported an oxygen flux through silver of 2.22×10^{-7} $cm^3/s/cm^2$ at 410°C at a driving force of 24 $Torr^{0.5}$ (nominal pressure of 576 Torr) for a Ag thickness of 200 µm. This corresponds to approx. $6.3.10^{-14}$ $mol/m/s/Pa^{0.5}$ of O_2, which is, at least, three orders of magnitude smaller than that which Pd-based membranes deliver in hydrogen. Under similar conditions, Cole obtained 3.21×10^{-7} $cm^3/s/cm^2$, which is in line with Gryaznov's values. Saracco and Specchia reported permeabilities of O_2 through Ag of 1.7×10^{-13} $mol/m/s/Pa^{0.5}$ and 6.3×10^{-11} $mol/m/s/Pa^{0.5}$ at 400°C and 800°C, respectively [60]. This involved an activation energy for oxygen permeation of approx. 90 kJ/mol, about 7 times more than that for H_2 permeation through Pd. The results from Saracco and Specchia differ from those of Gryaznov by 1 order of magnitude. The 2 data results provided by Saracco and Specchia contain, overall, greater imprecisions in comparison to the more numerous and detailed experimental measurements from Gryaznov, which have, in addition, been correlated by others [47]. The use of silver as an O_2 selective material in this work will require a strict adjustment of the process parameters (overpressure, layer thickness) in order to reach equivalent permeance with both selective membranes. In terms of the permeation mechanism of oxygen through silver, oxygen dissociates on the silver surface and diffuses through the bulk in atomic form, similarly to hydrogen through Pd [48]. Very few articles have been published up to now on the use of silver as a selective layer for oxygen in comparison to the hundreds of papers related to Pd and hydrogen. This imbalance makes the use of a silver membrane in the project more uncertain, in terms of the achievable results, when compared to the established hydrogen permeable Pd-based membranes. Based on Gryaznov results, the oxygen flux through 4 µm Ag at 150°C and 0.35 bar driving force would be approx. 5.7×10^{-8} mol/m^2/s. Under the same conditions, 3.0×10^{-2} mol/m^2/s of H_2 would be obtained through PdCu (difference of 5 orders of magnitude). To balance out this gap and achieve similar partial pressures for both the H_2 and O_2 in the reactor, the Ag layer will have to be much thinner than the one of the Pd-based membrane and subjected to a larger driving force in the case of a perfect defect-free layer (a molecular flow through defects would increase the permeance but reduce the selectivity of the film).

Porous membranes

The use of non-gas selective porous membranes is considered as an alternative to silver, which could present a too low permeability for the set of process parameters utilized in the respective experiments. Porous membranes based on sintered ceramic particles, which feature micrometric pores, are often used as a support for thin dense metal films. They can also be further coated by sol-gel methods to reduce the pore size and synthesize a UF-membrane (pore size ranging from 10 to 100 nm) or NF-membrane (pore size between 1 and 10 nm) [57-59]. These membranes can be used for gas dosage purposes but are not suitable for gas separation due to their large pore size vis-à-vis the kinetic diameter of gas molecules. The gas permeability of a porous membrane is difficult to evaluate as said membranes are usually employed for liquid filtration purposes; the permeation data provided by manufacturers is thus usually only valid for liquids.

An example in this connection would be the TRUMEM® membrane. This support is manufactured in Russia and offers TiO_2/stainless steel porous membranes with an average pore size of 0.2 µm featuring a water flux of 1100 $L/h/m^2$ at a water pressure of 1 bar. When an additional NF layer is applied to such a support by sol-gel methods, which thereby reduces the pore size down to 1 nm, the membrane obtained presents a water flow of approx. 120 $L/h/m^2$ at a water pressure of 1 bar [58]. One has to be aware that the air flow through a porous membrane can be 2 to 3 orders of magnitude higher than that of the water through the same porous media. In this case, a porous UF membrane with an average pore size of 50 nm would deliver an air flow of approx. 0.5 $mol/m^2/s$ at 1 bar of overpressure.

3. Development of a new reactor concept for improved benzene hydroxylation

3.1. Analysis of the common reactor configuration for benzene hydroxylation

Sato et al. showed in their study on the direct hydroxylation of methyl benzoate to methyl salicylate, that the hydrogen to oxygen concentration ratio in the reactor plays a crucial role in benzene hydroxylation [27]. Using a tubular membrane reactor, they noticed strong variations in the oxygen and hydrogen concentrations along the reactor axis. Where oxygen was fed together with benzene at the reactor entrance and hydrogen was dosed through the outer shell of the membrane tube, it was observed that oxygen was present in a large amount in the first part of the reactor and was fully consumed upon reaching the reactor end in the case of a moderate oxygen concentration in the feed, while hydrogen was missing at the reactor entrance but became dominant at the reactor end. These opposite gas concentration profiles were related to the reactor design, which enables a controlled dosage of one of the reaction gases in the system. Sato suggested that, in this configuration, hydroxylation might only occur in a narrow reactor area where an adequate concentration of both H_2 and O_2 exist [27]. In the entrance part of the reactor, the high oxygen content favored oxidation reactions leading to the formation of CO_2. In the central part of the reactor, oxygen and hydrogen were found in similar amounts, which contributed to the formation of active surface radicals. Phenol (benzene hydroxylation product) was obtained in the central area of the reactor. Finally, benzene hydrogenation products, such as cyclohexane and cyclohexanone, were also detected. Hydrogenation products were produced near the reactor end where hydrogen was present in large amounts whereas oxygen was nearly totally consumed.

Fig. 3.1-1 illustrates a tubular membrane reactor, as was used at the AIST and in the Netherlands to perform the Pd membrane-assisted gas phase benzene conversion to phenol.

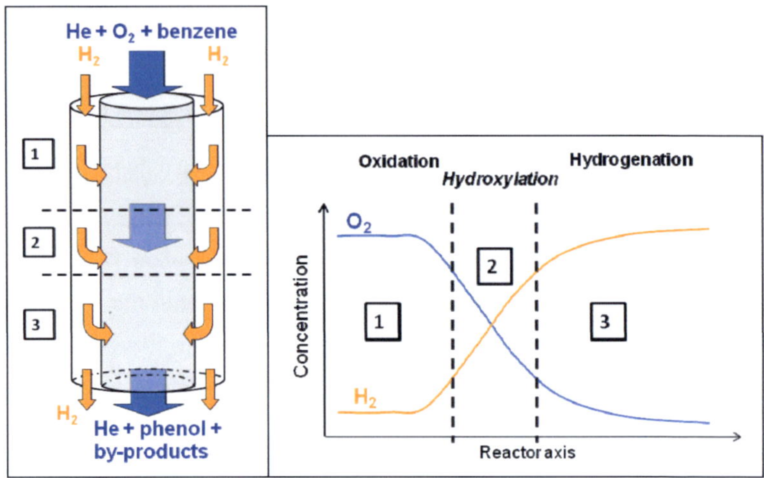

Fig. 3.1-1: Diagram of a tubular single membrane reactor, as employed by Sato et. al, and description of the dominant reactions with respect to the reactor part

The different reactor areas with the corresponding dominant reactions are described. They are determined by the hydrogen to oxygen concentration ratio along the reactor length.

As mentioned by Sato, the use of a tubular membrane reactor does not allow for an optimal use of the reactor volume, as only a narrow range of it is efficiently used to produce phenol. Phenol is obtained together with oxidation and hydrogenation by-products in the reactor.

Fig. 3.1-2 shows the experimental flow rates of the components H_2 and O_2 (consumption) as well as water (formation) presented by Sato et. al in the case of the direct hydroxylation of methyl benzoate to methyl salicylate [27]. According to the authors, this reaction involves many side reactions due to the nature of the reactant (decomposition to benzene) and atmosphere changes inside the reactor, namely from oxidative to hydrogenative conditions along the reactor axis.

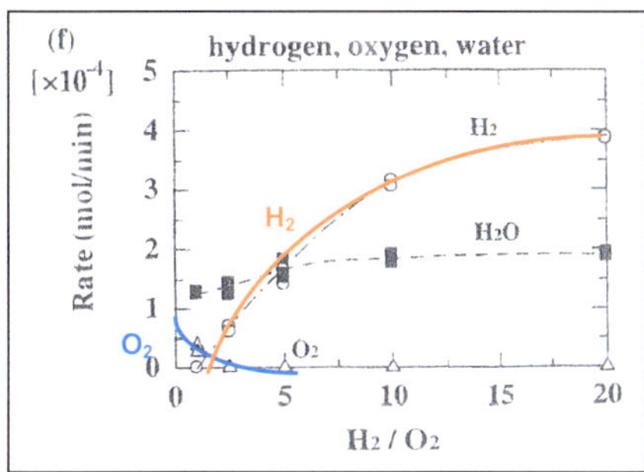

Fig. 3.1-2: Flow rate evolution of oxygen, hydrogen and water rates inside the reactor from Sato et. al in the case of methyl benzoate direct hydroxylation [27]. They are plotted vs. the H_2/O_2 ratio as it varies from the reactor entrance to the reactor end under the effect of the different reactions involved.

The diagram shows opposite concentration profiles of oxygen and hydrogen inside the reactor and is obtained when only hydrogen is dosed through a membrane. This results in different areas where oxidation, hydroxylation and hydrogenation are successively favored along the axis, which, in turn, gives rise to different types of products. The authors, however, do not state how the concentrations and the H_2/O_2 ratio were measured inside the reactor during the experiment (if said is technically possible). It is not clear whether the rates were measured or estimated, this does not allow for a concrete statement to be made on the reliability of the results in Fig. 3.1-2. The main idea behind these results is to illustrate the different concentration profiles of hydrogen and oxygen using 1 membrane, which limits the area for hydroxylation in the reactor.

3.2. New reactor design and realization

In this project, an improvement of the reaction conditions for phenol production by means of a different reactor design is proposed. A way had to be found to try to reduce the influence of side-reactions along the reactor length. With this objective, the idea was put forward to introduce a second membrane in order to also dose oxygen in the reactor and to achieve similar concentration profiles for both hydrogen and oxygen, which would lead to a stable H_2/O_2 ratio all along the reactor. By setting this ratio for phenol formation conditions, it would be possible to extend the hydroxylation area to the whole reactor range and thus improve the phenol selectivity of the system. The tubular design was changed to a flat design in order to make the reactor construction less challenging in terms of integrating both membranes (main difficulty: achieve sealing with 3 tubes). The flat design allows for the use of 2 flat membranes of similar surface, which face each other in the reactor. The new reactor design is illustrated in Fig. 3.2-1.

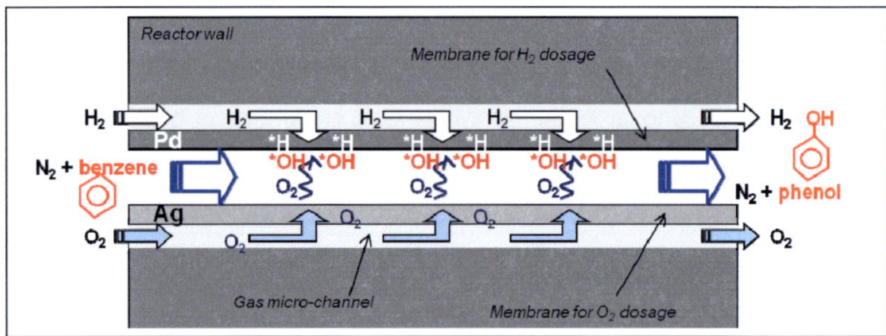

Fig. 3.2-1: Diagram of the new reactor design: reactor composed of two stainless steel plates supporting two membranes, which are positioned opposite to each other, for the separate dosage of both H_2 and O_2 in the reaction area.

Two stainless steel reactor plates, 18 cm x 7 cm, support the two membranes, the aim here being the separate dosage of hydrogen and oxygen in the central reaction channel. Each reactor plate is engraved with gas micro-channels to allow gas supply, at the retentate side of the flat membranes, along the reactor axis. The dimensions of the flat membranes are 5 cm x 14 cm; their edges are supported by the plates and covered on the other side by the flat sealing. The micro-channels extend over an area, which measures 12 cm (length) x 4 cm (width). They are parallel to each other and equally distributed underneath the membranes. The 39 micro-channels are V-shaped, with a surface width of 200 µm and a maximum depth of 200 to 220 µm. They have been engraved using a Nd:YAG laser – Raymarker 2000 from the company Laser Pluss AG (Kirschweiler, Germany) with adjusted parameters (see Appendix 1 for details).

A flat membrane with a selective layer based on palladium is used for the hydrogen dosage and PdCu is the preferred alloy as explained in the previous section. A selective metal layer based on Ag silver was the preferred choice for the oxygen dosage.

A microstructured aluminum plate machined with 26 reaction channels is positioned inside the reactor, in the area between the membranes, in order to guide the benzene flow along the reactor length. The distance between the two membranes is set by the flat sealing, 2 mm thick, and by the central plate, which is located in-between the two membranes. The reactor volume available for the reaction is estimated to be around 8 mL based on the part of the reactor space filled by the microstructured plate. The membrane area available for the reaction is approx. 48 cm^2.

Both plates are pressed together, the flat sealing (graphite or polymer sealing) and the other elements can be found in-between said plates. The reactor structure is secured with screws at the periphery and enclosed in the heating band. The reactor is then integrated in the experimental setup. It has 3 gas inlets and 3 gas outlets for the gas dosage in the central part and the reaction itself. In the upper reactor part, a gas inlet and outlet are employed for the hydrogen supply at the retentate side of the upper membrane, while those in the lower reactor part are used for the oxygen supply at the retentate side of the lower membrane. Those connected to the central reactor part are employed for the benzene introduction in the central reaction channel by means of a carrier gas (nitrogen).

The photos in Fig. 3.2-2 show the reactor elements separately (reactor plates, double membrane, central microstructured plate and sealing) and then assembled in the system. A heating band is used to set the reactor temperature, which will range from 150°C to 250°C, for the benzene hydroxylation reaction.

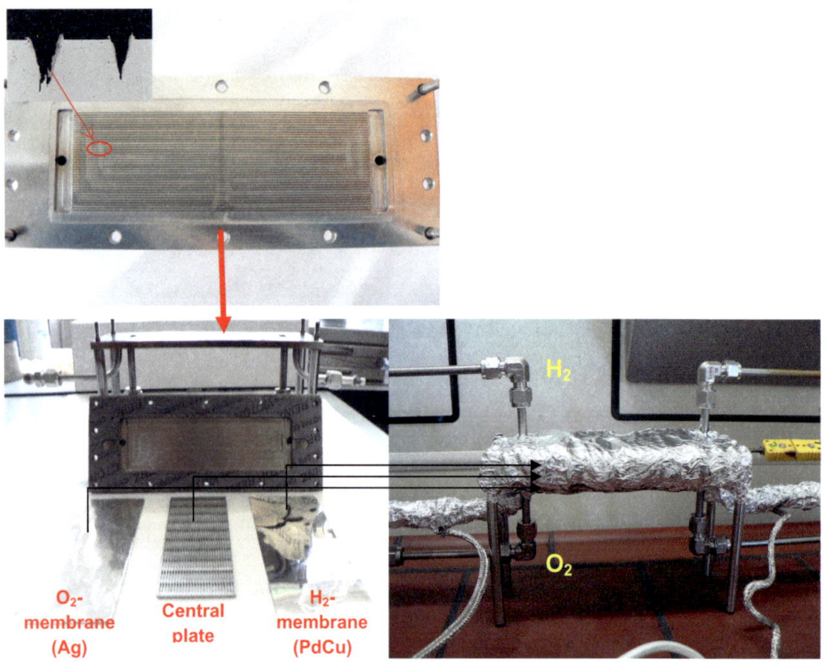

Fig. 3.2-2: Photos of the reactor assembly with views on the reactor plates, internal components (membrane foils and microstructured plate) and assembled reactor

3.3 Work modes of double-membrane reactor

3.3.1. Hydrogen/oxygen mixing possibilities

The use of 2 separate membranes for the dosage of oxygen and hydrogen in the system all along the reaction channel appears to be the best method to achieve a good control of the atmosphere composition from the reactor entrance to its outlet. This is necessary to reduce the influence of side-reactions and broaden the hydroxylation area to the whole reactor. This way of designing the reactor also allows for the possibility to use it as follows: to mix both reactive gases together with benzene in the feed as well as to "switch off" one or both of the membranes depending on the experimental conditions one wants to achieve. For example, the amplitude of the oxygen and hydrogen partial pressure is restricted by the membrane dosage and in particular the membrane thickness, it is possible for kinetic determination purposes to hereby dose a much larger amount of both gases, introducing them in the feed together with benzene and nitrogen.

Mixing together hydrogen and oxygen in the feed, however, raises safety issues at a wide partial pressure range. One has to be aware of the explosion limits of the H_2/O_2 and H_2/air mixtures depending on the process condition. The explosive domain of hydrogen in oxygen ranges from 4% vol. up to 94% vol. at atmospheric pressure; that of hydrogen in air extends from 4% vol. to 75% vol. at the same pressure. The ignition temperature for a spontaneous explosion of the mixture is about 560°C, which is much higher than the temperature range planned for this study. Precautions thus have to be taken when mixing together oxygen and hydrogen at the reactor entrance.

3.3.2. Dosage configurations depending on the membrane use

Due to the modularity of the reactor, 3 different work modes can be applied:

- *Double-membrane dosage mode of both H_2 and O_2 via the membranes*:

This is the usual work mode of the membrane reactor, where benzene is fed at the reactor entrance with a nitrogen carrier flow and the two reactive gases are dosed separately along the reactor length via the membranes positioned opposite to each other.

The major advantage of this configuration is that a quasi-constant concentration ratio of H_2 to O_2 should be maintained in the reactor from its entrance to exit, increasing the utilized area for benzene hydroxylation. Moreover, safe conditions are maintained in the system as the membranes prevent an uncontrolled direct mixing of oxygen in hydrogen, which could lead to the formation of an explosive atmosphere. The drawback in dosing gases through membranes is the limitation in the partial pressure level that can be reached in the reaction channel, as the entry flow can be more than an order of magnitude higher than the diffusive flow through the membranes (as the atomic diffusion through a compact layer is much slower than a convective flow). Therefore, neither high hydrogen nor high oxygen partial pressures can be reached, in this case, in the system. The partial pressures of both reactive gases are usually restricted to below 10% of the total pressure of the system due to the ratio of the membrane diffusive flow to the central channel convective flow. This mode with low amplitude of the H_2 and O_2 partial pressures will restrict the phenol formation kinetics. Higher kinetics would be achieved by a direct mixing of H_2 and O_2 in the reaction area. To enable such measurements, which may be of interest for kinetic analysis purposes, other gas dosage modes can be used, as described below.

- Mono-membrane dosage mode with one of the gases being co-fed with benzene at the reactor entrance:

One of the membrane channels can be left idle, while the opposite membrane is employed to dose one of the reactive gases. In this case, the safety condition will be guaranteed via a membrane separation. One of the partial pressures can also be monitored in a higher amplitude level via the direct introduction of the gas at the entrance, while the other one is dosed through the membrane. In this configuration, it will probably not be possible to maintain the homogeneous mixture of H_2 and O_2 in the long-term along the reactor; the different dosage mode will create a concentration gradient along the reactor as reported by Niwa et al. [25]. This should lead to a degradation of the phenol production rate when compared to the double-membrane dosage mode, due to a limitation of the active surface for benzene hydroxylation. However, this configuration enables a comparison between the double-membrane dosage and the literature results where only hydrogen was dosed through the Pd membrane and oxygen was introduced together with benzene at the reactor entrance.

- Co-feed dosage mode of both gases H_2 and O_2 together with benzene:

In this case, both membrane inlets are not utilized and all gases are mixed together at the reactor entrance. This configuration enables a higher variation of the oxygen and hydrogen partial pressures in the system, which facilitates kinetic measurements; however safety issues are raised due to the direct mixing of the hydrogen and oxygen when it enters the explosion range of the mixture (4 to 95 vol.% of H_2 in O_2). The low flows used in the membrane reactor during experimental measurements, kept below 400 mL/min, and the reaction temperature range (150-250°C), below the ignition temperature for explosion (560°C at 1 bar), result in a limited risk while using the reactor in this configuration. The formation of active radicals for hydroxylation will still occur on the Pd-based surface after the diffusion of gas H_2 and O_2 to the surface on the same side of the membrane. Only the decomposition property of the catalytic surface is used in this mode. A smaller number of experiments aimed at determining reaction kinetics will be performed in co-feed mode.

4. Experimental setup

4.1. Composition and realization

Fig. 4.1-1 illustrates the experimental setup for benzene hydroxylation. It was designed to make the collection and analysis of products in different phases possible: gas, liquid and solid products (benzene and phenol at room temperature, respectively). The experimental setup is made up of three main parts: benzene supply, phenol reaction and products recovery section.

- In the first section, benzene located in a saturator is introduced in the reactor via a nitrogen flow. The temperature applied to the saturator outer surface controls the benzene partial pressure on the inside (i.e. 0.04 bar at 0°C or 0.14 bar at 20°C). Nitrogen flow bubbles through the liquid benzene at a set flow rate and transports the corresponding benzene fraction through the heated lines as far as the reactor entrance.

- In the reactor, hydrogen and oxygen are dosed in the reaction area by means of the pressure difference set between the retentate and permeate sides of the respective selective membranes. Synthetic air is used as the oxygen source; oxygen is subsequently extracted and dosed in the reactor via the Ag membrane. The pressure at the retentate sides of the membranes is controlled by pressure controllers, which range from 1 to 3 bar. In the case of the porous membrane, very low overpressure is applied. The total pressure in the central part of the reactor is kept constant at 1 bar. The reactor temperature influences the gas dosage in the reactor. The temperature set for the reaction was varied between 150°C and 250°C.

- The organic products: phenol, unreacted benzene plus water are condensed in cooling traps and analyzed offline in a GC-MS system (Shimadzu QP5050). Two cooling traps can be alternatively used to collect the reaction products. This flexibility allows, for instance, for the separation of the products formed during start-up (high amount of water) until a steady state is reached in the reaction or for the extraction of a product sample while the reaction is in progress. The tubes from the benzene saturator to the cooling traps are heated in order to prevent condensation. The remaining gas phase is analyzed online in a micro gas chromatograph (Agilent 3000A). N_2, unreacted O_2 and H_2, plus potentially generated gases, such as CO_2 or CO, are detected with high accuracy in the two columns of the µGC.

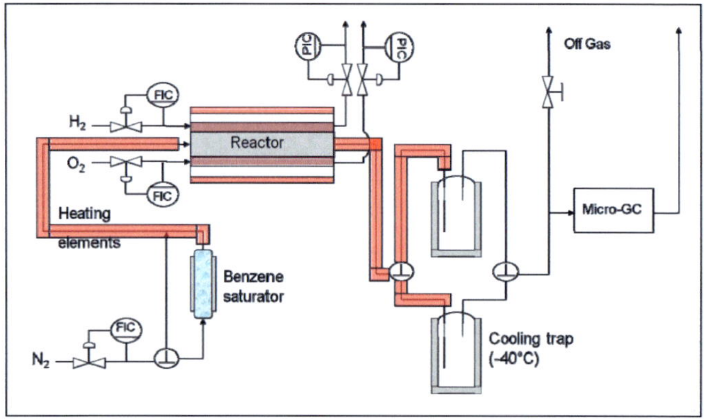

Fig. 4.1-1: Diagram of the experimental setup for benzene direct hydroxylation. The organic products are condensed in cooling traps and analyzed offline in a GC-MS system. The remaining gas phase is analyzed online in a 2 column micro-GC (Agilent 3000A for the detection of N_2, H_2, O_2, CO, CO_2 and organic gases up to C4).

Fig. 4.1-2 shows the realization of the experimental setup used for benzene hydroxylation with a view on the double-membrane reactor.

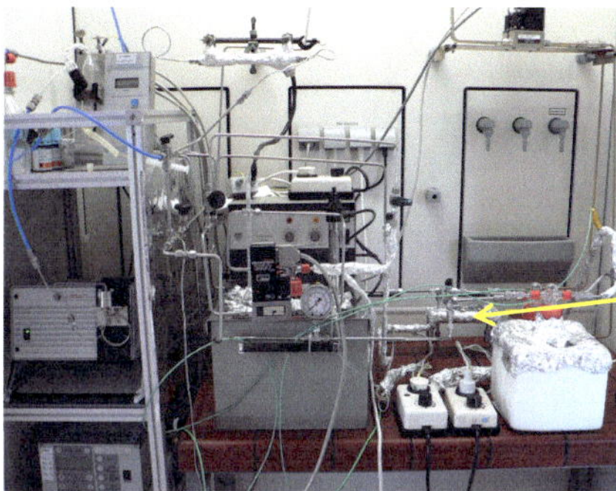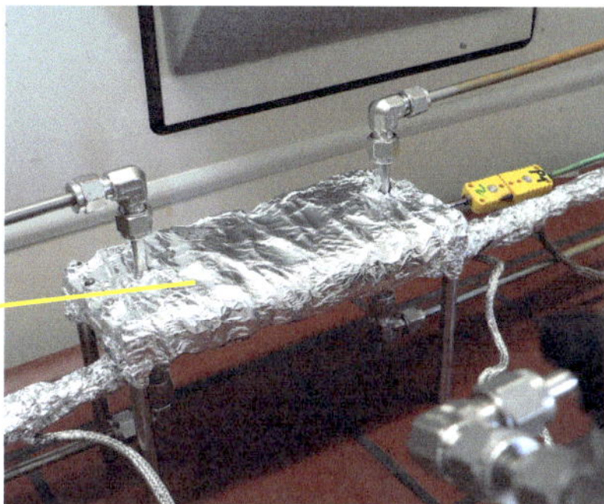

Fig. 4.1-2: Views of the experimental setup, including the benzene evaporator, reactor, heating band control system, cooling trap, MFC and PCs and micro-gas chromatograph

The parameters controlled in the process are the following:
- Nitrogen, synthetic air and hydrogen flows, ranging from 0 to 400 mL/min
- Hydrogen and synthetic air pressure, ranging from 1 to 3 bar
- Reactor temperature, ranging from 150°C to 250°C [25, 26]
- Membrane types and properties, which are employed in the reactor for the hydrogen and oxygen dosage and as a catalytic surface in the case of the Pd-based membrane. Dense metal foils and plated composite membranes were both examined as part of the scope of this work.

The required temperature for the cooling trap was determined by simulation using the Aspen Plus. The system was simulated with a focus on the phase equilibrium in the cooling container. Assumptions on the reaction rates and process conditions (entrance flows and temperatures) had to be made for the prediction of the product composition and flow entering the cooling container. Fig. 4.1-3 depicts the results of the simulation, giving the gas phase fraction of organic components in the cooling container according to the temperature. It is observed, for example, that phenol is fully recovered in the cooling trap for a temperature of 0°C, whereas a large fraction of unreacted benzene, close to 30%, remains in the gas phase at the same temperature. To reduce the benzene loss to 3%, a temperature of -40°C has to be reached in the cooling trap. This value was set as the target temperature, which was reached using a mixture of $CaCl_2$ and ice in the cooling trap in a weight ratio for calcium chloride/ice of 1.22. The cooling trap was isolated from the laboratory atmosphere via a polystyrene shell. The temperature in the trap was measured over time and deviations from the target temperature of -40°C were balanced out by small additions of calcium chloride.

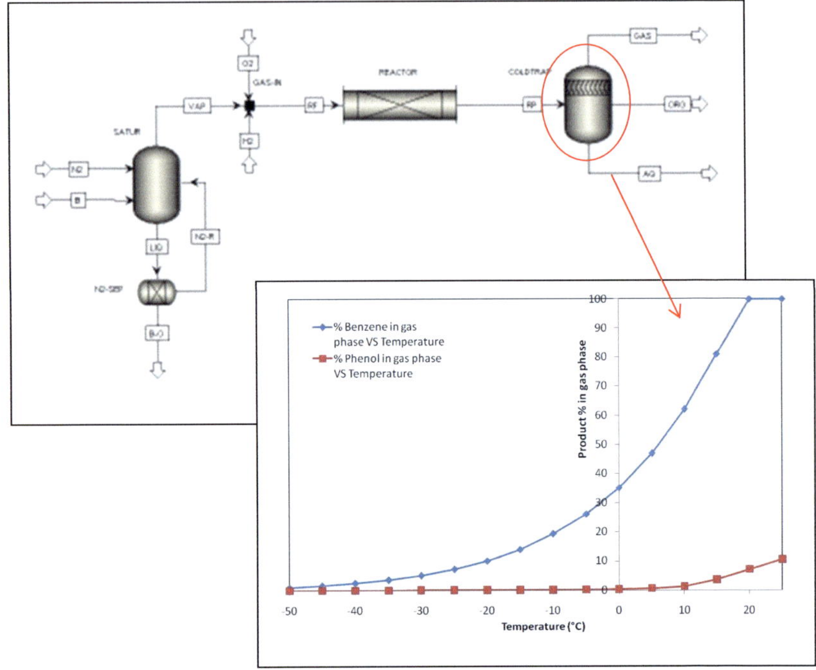

Fig. 4.1-3: Simulated loss of benzene and phenol in the cooling trap as a function of the cooling trap temperature (Aspen Plus).

Due to the fact that benzene cannot even be completely recovered at -40°C, the benzene conversion was calculated from the C-material balance, taking into account the sum of all measured products and the initial benzene amount dosed in the system, instead of considering the measured benzene amount.

4.2. Experimental procedure

The reactor assembled with the chosen membranes for hydrogen and oxygen dosage was placed in the setup and connected to the benzene nitrogen feed channel and hydrogen and oxygen retentate channels via Swagelok fittings (see Fig. 4.1-2).

A feed of pure nitrogen was used before the start of the experiments to sweep the reaction channels while the reactor and gas tubing was being heated. Then, hydrogen and oxygen were fed into the retentate channels of the membranes and an over-pressure was applied to initiate permeation into the central reaction channel, which was kept at constant atmospheric pressure. The reaction temperature, feed flow and pressure gradient were set according to the desired parameter values. Once the H_2/O_2 flows reached a steady state – this usually takes a few minutes – (control via online gas sample extraction and analysis with the Agilent 3000A µGC), the feed was switched to nitrogen and benzene and the cooling trap for sample collection was connected at the reactor end. The reaction investigation started at this point. While the reaction products were condensing into the cooling trap at -40°C, a sample of the residual gas phase was extracted and analyzed every 10 minutes. The reaction temperature was permanently checked together with the pressure gradient applied between the 2 membranes and the central channel. After several hours (for instance 3 or 4 hours if the preparation, experiment run and analysis should take place on the one working day), the reaction was stopped and the cooling container was extracted for product analysis (see Fig. 4.2-1).

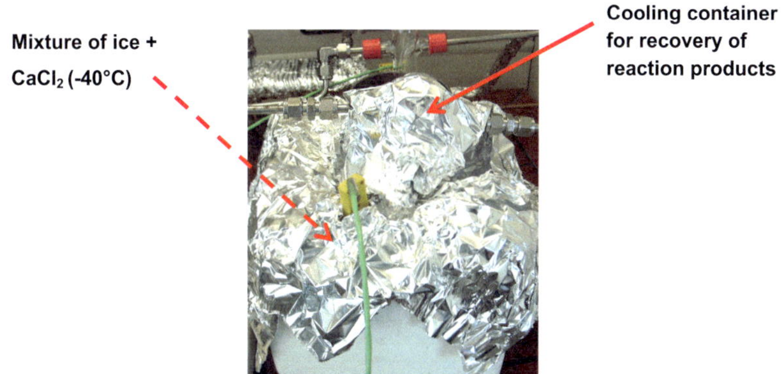

Fig. 4.2-1: View of the cooling trap for product recovery.

The cooling container was weighed to obtain the total product mass (compared to the weight of the empty container). Then the container was opened and the organic part of the frozen mass (-40°C) was dissolved in 10 mL of pentane (organic solvent), transferred into a 100 mL flask, topped up to 100 mL with more pentane and homogenized. The container with the undissolved aqueous product "clinging" to the wall (water produced from the reactions involved) was weighed again to determine the water amount. 10 mL of the organic product solution was then extracted from the 100 mL flask and introduced into 2.5 mL sample glasses for GC-MS analysis. Multiple GC-MS analyses gave the nature and amount of the products (unreacted benzene, phenol, etc...) found in the 2.5 mL glasses. These were then converted into the corresponding amounts contained in the full 100 mL solution. All product amounts (solid, liquid and gas) were expressed in mol and divided by the experiment duration for production rate comparison.

As an example, Fig. 4.2-2 illustrates the gas partial pressures measured at the reactor end with the Agilent 3000A micro gas chromatograph. N_2 (carrier gas), O_2 and H_2 (reactive gases) as well as CO_2 (product of oxidation reaction) were detected.

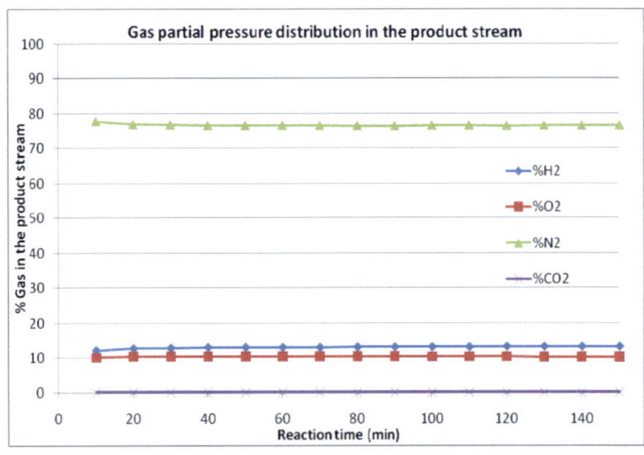

Fig. 4.2-2: Gas partial pressures measured at the reactor exit (behind the cooling trap) during an experiment performed at a set H_2/O_2 ratio.

4.3. Expression of the reaction performance

The following equations are used for calculating the performance:

Educt conversion (%):

$$X_i = \frac{n_{i0} - n_i}{n_{i0}}$$

With n_{i0} the initial mole flow of educt i introduced into the reactor and n_i its mole flow at the reactor end. With regard to the benzene consumption in the reactor, one can express the benzene conversion as the ratio between the benzene consumed and the benzene dosed in the reactor. This is also the ratio between the sum of the product rates based on benzene and the initial benzene rate:

$$X_B = \frac{\sum\limits_{k} \frac{1}{\left| v_{k,B} \right|} r_{k,B}}{r_B}$$

With $r_{k,B}$ the formation rate of the product k based on benzene, $v_{k,B}$ its stoichiometric coefficient, and r_B the benzene feed rate in the system (mol/h). Using the same principle, the hydrogen and oxygen-based conversions are the following respectively:

$$X_{H2} = \frac{\sum\limits_{k} \frac{1}{\left| v_{k,H2} \right|} r_{k,H2}}{r_{H2}} \quad \text{and} \quad X_{O2} = \frac{\sum\limits_{k} \frac{1}{\left| v_{k,O2} \right|} r_{k,O2}}{r_{O2}}$$

With r_{H2} and r_{O2} the hydrogen and oxygen feed rates in the system, respectively, and $r_{k,H2}$ and $r_{k,O2}$ the rates of the products based on H_2 and O_2.

Product yield at the reactor end (%):

$$Y_k = \frac{n_k - n_{k0}}{n_{i0}} \cdot \frac{\left| v_i \right|}{\left| v_k \right|}$$

Where Y_k is the yield of the product k, n_k the mole flow of the product formed, n_{k0} its initial mole flow in the system (if present), v_k the stoichiometric coefficient of the product k, n_{i0} the initial mole flow of the educt i from which k is formed and v_i the stoichiometric coefficient of the educt i. Based on the formation rates, the yield of the product k created in the reactor can be described by:

$$Y_k = \frac{r_k}{r_{i0}} \cdot \frac{|v_i|}{|v_k|}$$

Product selectivity at the reactor end (%):

$$S_{k,i} = \frac{Y_k}{X_i}$$

The selectivity of the product k formed from i is defined by the ratio between the yield of k and the conversion of i. For instance, the carbon-based selectivity (with benzene as the educt) can be written as follows:

C-based selectivity (%):

$$S_{k,B} = \frac{\dfrac{r_k}{|v_k|}}{\sum_k \dfrac{1}{|v_{k,B}|} r_{k,B}}$$

The hydrogen and oxygen based-selectivities can be described as follows:

H-based selectivity (%):

$$S_{k,H} = \frac{\dfrac{r_k}{|v_k|}}{\sum_k \dfrac{1}{|v_{k,H}|} r_{k,H}} \qquad \text{with k the products based on H}$$

O-based selectivity (%):

$$S_{k,O} = \frac{\dfrac{r_k}{|v_k|}}{\displaystyle\sum_k \frac{1}{|v_{k,O}|} r_{k,O}}$$

with k the products based on O

With a special focus on phenol, the expected reaction product, it is possible to express its productivity as the ratio of the phenol formation rate and the catalyst amount or catalyst surface area:

$$P_{phenol} = \frac{r_{phenol}}{m_{cat}}$$

It is, however, difficult to establish the effective catalyst amount for the reaction, as only the permeate surface of the Pd-based membrane acts as a catalytic surface, while the palladium also constitutes the bulk of the membrane for hydrogen dissociation and transport purposes. The situation is further complicated due to the fact that the catalytic surface was also modified in this project (modified composition and increased thickness). For this reason, it is preferable to consider the *phenol rate in mol/h* as the variable for describing the reactor performance under certain process conditions, together with the *phenol selectivity* and *benzene conversion*, both in %.

5. Analysis techniques

5.1. Optical Microscopy

Optical microscopy uses visible light, typically coming from an artificial light source, to amplify the features of a sample's surface. A beam of light is focused onto a tiny spot of the sample's surface via a system of lenses called the condenser. The image of the object is brought into focus via the microscope tube by means of a spherical objective lens. A second ocular lens then magnifies the image of the object. The magnification of the analyzed area is increased by changing the objective lenses: relatively flat lenses produce low magnification in comparison to more spherical lenses. The quality of the image is determined by the brightness, focus, resolution and contrast. The brightness depends on the illumination system and can be adapted by changing the lamp voltage (rheostat) and adjusting the condenser and/or diaphragm apertures.

The focus is related to the focal length and is controlled using focus switches. The capacity to distinguish two neighboring points is known as the resolution. This is related to the numerical aperture of the objective lens and the wavelength of the light passing through the lens. A better resolution is attained with a higher numerical aperture and a shorter wavelength. The minimal appreciable distance d_{min} can be calculated using the equation:

$$d_{min} = \frac{0.6\lambda}{n \sin \alpha}$$
Eq. (5.1-1)

with α the overture angle, λ the wavelength and n the refractive index. Optical microscopy can then distinguish objects separated down to *0.2* μm.

The contrast is related to the illumination system and can be adjusted by changing the intensity of the light and the diaphragm aperture. The contrast can be improved by closing the condenser aperture, but this tends to reduce the resolution.

5.2. Scanning Electron Microscopy with EDX analysis

5.2.1. Generalities and composition of a scanning electron microscope

The scanning electron microscope (SEM) images the respective sample surface by analyzing the signals emitted in the case of the electron-solid interaction. It is used to characterize solid topography, chemical bonds, atomic distance or polarization features. In combination with other experimental techniques, namely XRD and TEM, it constitutes a fundamental investigation technique in the fields of metallurgy, chemical engineering, electronics and geology.

The resolution capability of the SEM is much higher than that of optical microscopy since the wavelength of the electron beam used in the SEM is around 10^5 times smaller than that of light, making the resolution limit 10^5 smaller than that of an optical microscope (see Eq. 5.1-1).

A scanning electron microscope consists of a vacuum chamber, the electron microscope body and the sample carrier. The body parts are:

- the electron gun
- the lenses to focus the ray and for scanning the specimen surface
- the detector for the signals released from the specimen

as illustrated in Fig. 5.2-1, where the "virtual source" represents the cannon of electrons.

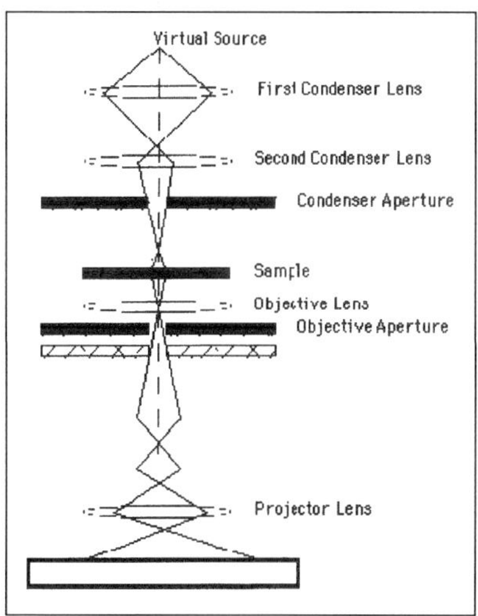

Fig. 5.2-1: Diagram of the components of a scanning electron microscope [56]

5.2.2. Working principle of a scanning electron microscope

In many microscopes, the pressure in the body is of the order 10^{-7} Pa. A beam of accelerated electrons in the desired potential can be obtained by means of the electron cannon. The emission is produced either by thermo-electric effect or field effect.

The thermo-electrical guns can actually be made from two materials: W or LaB_6. A standard tungsten wire suffices for the classic applications (resolution of 5 μm). A mono-crystal tip of LaB_6 is used (resolution 3.5 μm) where a high resolution is necessary; the constraints then, however, become more important, < 10^{-7} Torr (1 Torr = 133.3 Pa). The tungsten capillary-cathode is normally selected as the thermal cathode. A bent tungsten wire (filament), fastened on a ceramic base, can be stimulated by applying a heating voltage for the thermal emission of the electrons. Temperatures between 2700 °C and 2800 °C are required in order to reach the working level of 4.5 eV; as it is only then that the electrons build a space charge in front of the cathode. The electron beam subsequently passes through a system containing the magnetic condenser and objective lenses, which allows for the focusing of the beam on the respective target. Diaphragms limit the opening for the beam; while its diameter is determined by the crossover reduction due to the lens system. It reduces the beam diameter to about 3-20 nm.

The primary electrons of the beam produce different interactions with the specimen, which are used to generate the image, namely released secondary electrons and backscattered electrons (material contrast). A primary electron colliding with the specimen can release up to four secondary electrons from the surface, otherwise it would be repelled by the specimen and detected as a backscattered electron. The electrons emitted by the sample reach an electron multiplier or fall on a light sensitive layer (ZnS), which produces photons. These photons are then guided towards a photo-multiplier by means of a light guide. The images generated using secondary electrons are essentially formed by low energy electrons.

A resolution, which is 30 times higher when compared to a light microscope (4 nm with tungsten capillary-cathode), and a much higher degree of sharpness are the big advantages of the scanning electron microscope. Limitations only arise in relation to the required vacuum and the specimen conductivity as without the latter it would be impossible to create an image. For this reason, carbon is often sputtered onto the surface of specimens in order to improve their conductivity.

5.2.3. Interactions between electron beam and sample

The volume inside the bulk specimen in which interactions with the electron beam take place is called the specimen interaction volume. This volume depends on the atomic number of the material being examined, the used accelerating voltage and the angle of incidence of the electron beam.

Figure 5.2-2 depicts the interactions of a specimen, whose main element has an atomic number of 28, with an electron beam, which has an accelerating voltage of 20 kV and 0 degree tilt.

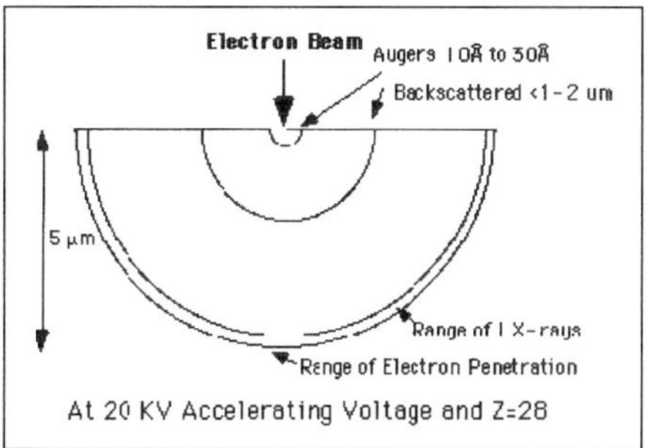

Fig. 5.2-2: Diagram of the components of an electron gun [56]

A number of interactions between the material and the electron beam take place in the interaction volume. The following points describe those interactions, which are used for the characterization of the material in SEM.

<u>Backscattered Electrons</u>: They are caused by an incident electron colliding with an atom, which is nearly perpendicular to the incident's path in the specimen. The incident electron is then scattered "backward" 180 degrees.
The production of backscattered electrons directly depends on the specimen's atomic number. These deviating production rates cause higher atomic number elements to appear brighter than lower atomic number elements. This interaction is utilized to differentiate parts of the specimen that have different average atomic numbers.

<u>Secondary Electrons</u>: They are caused by an incident electron passing "near" an atom in the specimen, near enough to transfer some of its energy to a lower energy electron (usually in the K-shell). This causes a slight energy loss and path change in the incident electron and the ionization of the atom. This electron then leaves the atom with a very small kinetic energy (5 eV) and is then termed a "secondary electron". Each incident electron can produce several secondary electrons. The production of secondary electrons is extensively related to the topography. Due to their low energy, only secondary electrons, which are max. 10 nm from the surface can exit the sample and be

examined. Any changes in the topography of the sample, which are larger than this sampling depth will change the yield of secondary electrons due to collection efficiencies.

<u>Auger Electrons</u>: They are caused by the energy loss of the specimen atom after a secondary electron is produced. Since a lower (usually K-shell) electron is emitted from the atom during the secondary electron process, an inner (lower energy) shell now has a vacancy. A higher energy electron from the same atom can "fall" to a lower energy, filling the vacancy. This creates an energy surplus in the atom, which can be corrected by emitting an outer (lower energy) electron; an Auger electron.

Auger electrons have a characteristic energy, unique to the element from which they were emitted. These electrons are collected and sorted according to their energy level in order to obtain compositional information about the specimen. Since Auger electrons have a relatively low energy, they are only emitted from the bulk specimen from a depth less than 3 nm.

<u>X-rays</u>: They are caused by the energy loss of the specimen atom after a secondary electron is produced. Since a lower electron, usually from the K-shell was emitted from the atom during the secondary electron process an inner (lower energy) shell now has a vacancy. A higher energy electron can fall into the lower energy shell, filling the vacancy. As the electron falls it emits energy, usually X-rays. X-rays emitted from the atom will have a characteristic energy which is unique to the element from which it originated. The elemental composition of the specimen can be measured as the energy of the X-rays is characteristic of the difference in energy between the two shells, and of the atomic structure of the element from which they are emitted.

EDX (Energy Dispersion X-Rays) measurements performed within the scanning electron microscope, where equipped with such a detection system, allow for the identification of the specimen elements, in addition to the surface analysis. The SEM/EDX combination is extensively used for sample characterization purposes.

The scanning electron microscope, used at the Karl-Winnacker Institute for sample analysis, in this work is a Philips XL40, equipped with an EDAX detection system.

5.3. Gas Chromatography and Gas Chromatography combined with Mass Spectroscopy

5.3.1. Gas Chromatography

To provide a brief historical background, the term chromatography can be traced back to the first decade of the 20th century, as this was when this method was first described by the Russian scientist, Mikhail Semenovich Tswett. The different variations of this analysis technique were subsequently developed in the 1940s by different scientists: liquid-liquid chromatography in 1941, solid state gas chromatography in 1947, liquid-gas chromatography in 1950.

Gas chromatography is an analytical method aimed at identifying and quantifying the gas phases present in a sample mixture via the separation of said in a heated column. A mobile phase transports the gas mixture in the gas chromatograph from the sample injector to the detector - travelling through the column (typically characterized by its length and internal diameter) - where it interacts with a stationary phase (mainly polymeric), which is either packed in the column or coated on the column wall. Due to their specific interactions with the stationary phase, each compound exits the column at a different time, this is known as the retention time. The measured retention time allows for the identification of the gaseous species, while its concentration is proportional to the intensity of the detected peak (electronic identification in the detector). The stationary phase acts as a separator vis-à-vis the mixture, which is transported in the column via an inert carrier gas. The motion of the sample molecules is, in fact, inhibited by their adsorption on the stationary phase as the carrier gas sweeps said molecules through the column. The rate at which the molecules progress along the column depends on the strength of the adsorption, which in turn depends on the type of molecule and the stationary phase materials. Each type of molecule has its proper rate of progression in a given type of column and will therefore be detected at a specific retention time. The column temperature and gas flow will also influence the retention time.

Fig. 5.3-1 presents a gas chromatograph, typically composed of a sample injector, column and detector.

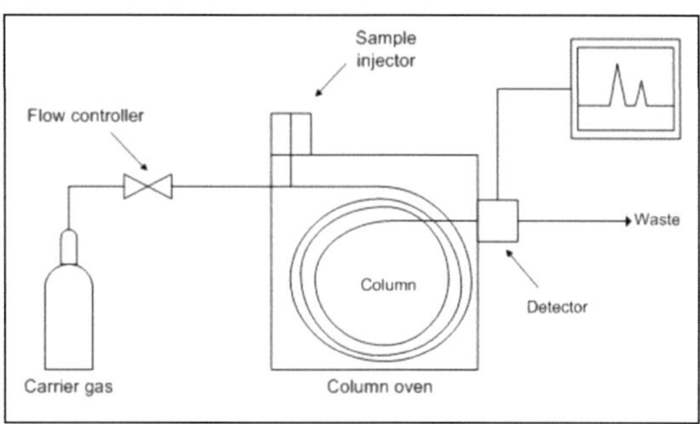

Fig. 5.3-1: Diagram of the gas chromatograph with its internal components and data acquisition system

<u>The sample injector (column inlet)</u>: Several types of injectors exist (split/splitless, PTV, etc.), recent variants are electromechanical systems equipped with a backflush head to prevent the intrusion of liquid phases in the column. A membrane extracts residual water vapor from the gas sample.

<u>The column</u>: Gas chromatographs can be equipped with up to 4 columns to ensure a high precision for the detection of a large range of gases and C-based gas phases. The most common types of columns are packed columns and capillary columns. Packed columns are usually 1.5 to 10 m long with internal diameters ranging from 2 to 4 mm and contain a packing of the stationary phase chosen for gas adsorption. Capillary columns have a very small internal diameter (sub mm domain) with a length up to 60 m. The interior walls are either coated with the active materials (WCOT) or solid-filled with a network of parallel micropores (PLOT). The system with the finest pores,

known as Molsieb columns, contain pores in the Å-domain that enable a molecular sieving of gas phases.

<u>The detector</u>: A large number of detector types are used in gas chromatography. The most common types are the flame ionization detector (FID) and the thermal conductivity detector (TCD). Both allow for the detection of a large range of components in a large range of concentrations. TCDs are non-destructive detectors, able to detect any component differing from the carrier gas in terms of thermal conductivity. FIDs are more sensitive to hydrocarbons than TCDs, however, they can only detect hydrocarbons. They fracture (destructive method) and ionize the molecules; the fragments are subsequently detected via their mass to charge ratio.

The gas chromatograph employed to perform the analysis of the gas phase in this work is a micro-gas chromatograph 3000A from Agilent. The injector has a volume of 1 µL and is equipped with a backflush system where a membrane eliminates humidity from the injected gas sample. The µGC 3000A is equipped with 2 columns delivering data on 2 channels: Channel A is a molecular sieve, which uses argon as the carrier gas, for the detection of H_2, O_2, N_2, CO, CH_4 and hydrocarbons up to C4. Channel B is a PLOTQ channel, which allows for a sensitive detection of CO_2 down to 0.1 ppm; in this case helium is the carrier gas. The detector has an internal volume of 240 nL and is based on TCD. Fig. 5.3-2 shows the transportable gas chromatograph with the data acquisition system and a diagram illustrating a signal obtained with channel A (molecular sieve).

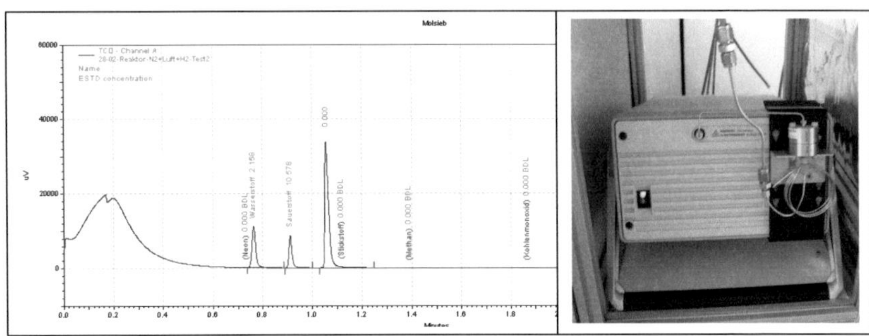

Fig. 5.3-2: View of the Agilent 3000A µ-GC with an example of signal detected on channel A (carrier gas: Ar)

5.3.2. Gas Chromatography combined to Mass Spectroscopy (GC-MS)

Gas chromatography-mass spectroscopy is an analysis method that combines the features of gas chromatography and mass spectroscopy to identify several substances dissolved in a volatile sample containing for instance liquid and dissolved solid phases when taken at room temperature. It is widely employed in analytical chemistry and several other fields requiring the detection of unknown samples, such as in environmental analysis, drug detection or explosives investigation.

The GC-MS is composed of two major parts: the gas chromatograph and the mass spectrometer. The gas chromatograph utilizes a capillary column, working on the same principles as the single GC instrument described above. The difference in the chemical properties of the gas species, which make up the mixture and are injected in the column, is responsible for the separation of the molecules as they travel in the column, leading to different retention times. As they exit the column separately, the online mass spectrometer captures, ionizes and detects the molecules flowing downstream. The mass spectrometer detects the ionized molecule fragments thanks to their mass to charge ratio and accordingly identifies them. The combination of both analysis techniques enables an accurate molecule identification of a particular molecule.

The samples are introduced via an inlet into the column, by means of a simple injection through a septum. Once injected, the heated chamber acts to volatilize the molecules to be identified; they are then transported into the column by means of an inert carrier gas. The detection is then realized by the mass spectrometer, which can be set to either full scan mode (identification among a wide range of m/z numbers) or selected ion monitoring mode when targeting a particular molecule or molecule group (when the sample composition is known).

Fig. 5.3-3 displays a GC-MS with its two online mounted blocks.

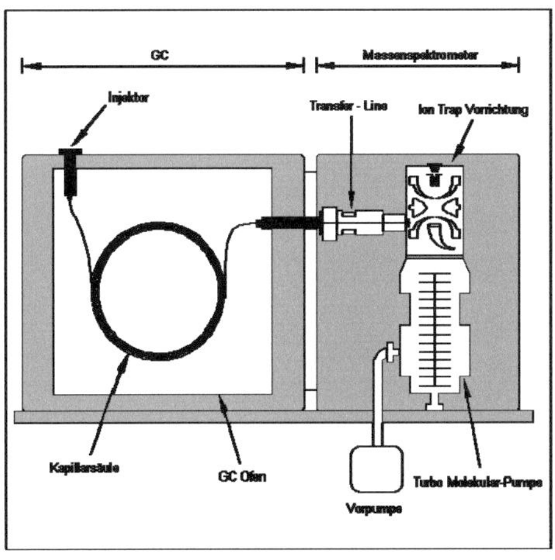

Fig. 5.3-3: Diagram of a GC-MS system showing the internal components

A GC-MS QP5050A system from Shimadzu was used for this work in order to identify and quantify the solid and liquid organic products generated by the double-membrane reactor (phenol, residual benzene, possible hydrogenation byproducts, etc...).

6. Membrane preparation and compositional analysis

6.1. Membrane selection

For comparison purposes, two types of membranes were tested in the reactor: dense self-supported metal foils and plated composite membranes produced by coating a commercial support with thin layers of Pd, Cu and Ag. The foils were manufactured and supplied by local companies. The 50 µm thick $Pd_{60}Cu_{40}$ wt.% foils from the company MaTeck (Jülich/Germany) were decided on for the hydrogen dosage, while the 15 µm thick silver foils from Schlenk (Roth-Bernlohe/Germany) were selected for oxygen dosage. The amount of gas dosed through the foils is controlled by the partial pressure difference between both sides of the membranes. It is known that silver has a much smaller oxygen permeability than palladium for hydrogen; the flux, in this connection, is estimated to be 10 times higher under the same conditions (T, P, membrane dimensions).

Composite membranes, which are composed of successive depositions of Pd and Cu layers and subsequently heat treated to form the PdCu alloy, were employed for hydrogen dosage [44-45; 53-55]. For oxygen dosage, single layers of silver were applied on the support in order to obtain the required selectivity. The layer deposition method, an electroless deposition process in the liquid phase, will be detailed in the next section. Electroless plating is a well-known method for forming, in particular, Pd-based composite membranes. It is also less cost-intensive than sputtering (no hardware required).

6.2. Common substrate for the composite membranes

The commercial support chosen for the composite membranes is a flat TRUMEM® membrane, suitable for liquid filtration, which was supplied by ASPECT (Moscow, Russia). It is composed of a porous stainless steel mesh covered by a thin porous TiO_2 layer, which acts as a diffusion barrier layer (see images in Fig. 6.2-1). The pore size present on the support is 0.2 µm. The used membrane surface is 14 cm x 5 cm, while the surface area accessible to the gas is 12 cm x 4 cm. The TRUMEM® was used as a support for the formation of both Pd/Cu and Ag layers in order to form the H_2 and O_2 selective membranes, respectively.

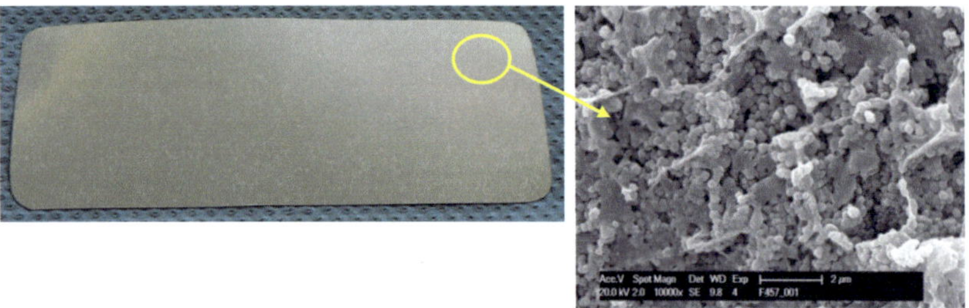

Fig. 6.2-1: View on the TiO$_2$ side of the TRUMEM® support used for the layer deposition and surface morphology of the porous TiO$_2$ layer (SEM picture)

6.3. Support activation method prior to Electroless Plating

The method used to produce the selective coatings on the support is based on an autocatalytic liquid phase reaction involving the reduction of metal ions from a plating solution on an active support, this leads to the deposition and growth of a metal layer on the support surface. To make this deposition and growth possible, the support has to be activated. If the support does not exhibit an adequate surface state (as is the case for the TiO$_2$ surface of the TRUMEM® support in its delivered state), then it must be activated prior to the electroless deposition by seeding the inert surface with palladium germs. Several methods may be used to achieve this activation.

6.3.1. Activation with SnCl$_2$/PdCl$_2$ baths

The most common activation method is a sensitizing of the support by successive immersions in tin chloride and palladium chloride baths in order to fix Pd nuclei onto its surface. This method requires the preparation of 2 sensitizing baths in which the sample, to be activated, will be successively immersed for 5 minutes and then rinsed with distilled water. The composition of both baths is detailed in Table 6.3-1.

Table 6.3-1: Composition of the $SnCl_2/PdCl_2$ activation baths

Bath 1: $SnCl_2$ activation bath	Bath 2: $PdCl_2$ activation bath
1 g $SnCl_2$ / L solution	0.2 g $PdCl_2$ / L solution
1 mL HCl	1 mL HCl
~1 L distilled H_2O	~1 L distilled H_2O

A sensitizing cycle involves the following: 5 min immersion in bath 1, followed by a brief cleaning of the surface with distilled water and another 5 min immersion in bath 2. The reaction taking place on the substrate surface is in line with Eq. 6.3-1:

$$Sn^{2+} + Pd^{2+} \rightarrow Sn^{4+} + Pd^0 \qquad\qquad \text{Eq. (6.3-1)}$$

A good activation of the support is achieved by repeating the activation cycle between 4 and 10 times. A compromise has to be found, depending on the substrate surface state, as too many sensitizing cycles produce a thicker activation layer, which lowers the adhesion of additionally deposited Pd layers on the support. This method was described by Ma and Mardilovitch [31-33] and is widely employed. A typical size for the Pd particles produced by this method and fixed on the surface is about 50 nm. As smaller Pd seeds cannot be obtained, the activation usually affects the surface and near sub-surface, depending on the pore size of the support. A deep anchorage of the Pd layer on the support is thus not achievable as the seeds cannot reach the finest pores below the surface.

If the Pd membrane is subjected to mechanical stresses and thus requires a higher adherence of the Pd coating, then an activation method, which supplies finer seeds is recommended. In the case of long-term applications of the Pd-membrane, it is suspected that the Sn species trapped at the interface with the

Pd layer (due to bad rinsing after activation) would result in the degradation of the membrane properties due to the ageing effect. For a laboratory use, with a restricted pressure gradient through the membrane (3 bar as maximum retentate pressure), no delamination of the coating or disadvantage in relation to the activation bath process has been observed.

The activation of the membrane support via $SnCl_2$ and $PdCl_2$ sensitizing baths was the preferred method in this case due to its good repeatability and convenience for activating flat surfaces. The support has, indeed, only to be sensitized on 1 side, which faces upwards in the bath containers. Membranes with a more complex geometry, such as a cylindrical form, would have to be rotated in the baths in order to ensure a homogeneous activation over the entire surface. In this case, the second activation method, using Pd-acetate, which is described below, might be more suitable. Fig. 6.2-2 shows the surface of a TRUMEM® membrane before and after activation via immersion in the $SnCl_2$ and $PdCl_2$ solutions. The change in surface appearance due to the presence of the Pd seeds is easily recognizable.

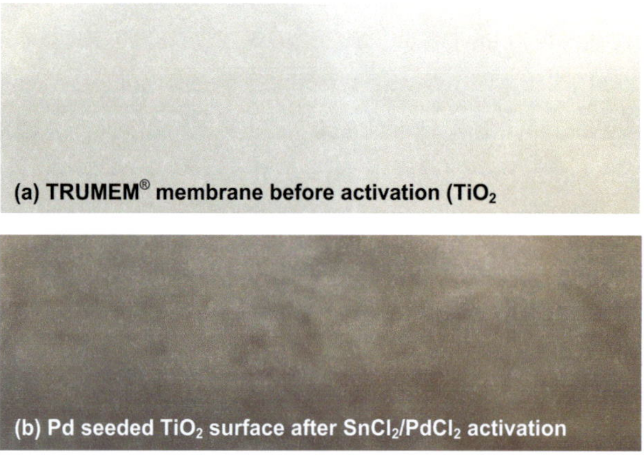

Fig. 6.2-2: View of the clean surface of the TRUMEM® membrane (TiO$_2$ face, pore size: 0.2 µm) before activation (a) and after activation using the

SnCl$_2$/PdCl$_2$ method (b) showing a darker appearance due to the presence of anchored Pd seeds

Both supports, used for the production of the PdCu and Ag membranes, were activated by seeding the surface with palladium particles via the SnCl$_2$/PdCl$_2$ method. The adhesion of the silver layer on the Pd-activated support is similar to the adhesion of the palladium film. The presence of Pd seeds, which do not form a film on the surface, rather a pattern of spaced particles, should have a negligible resistance effect on the oxygen transport through the silver layer.

6.3.2. Activation by thermal decomposition of Pd-II-Acetate

This method, which was developed in-house involves:
- the dissolution of Pd-II-Acetate crystals in acetone,
- the impregnation of the surface to be activated with this solution followed by the rapid evaporation of the acetone,
- the thermal decomposition of Pd-II-Acetate located on the support surface and in the pores to metallic Pd, at 250°C with air for a few hours.

With this method, finely dispersed Pd particles of about 10 nm are produced; they are distributed all over the support as well as in the pores located below the surface. Large amounts of Pd-II-Acetate can reach the small pores inside the support during the impregnation due to the high solubility of the Pd-II-Acetate in the acetone, in comparison to PdCl$_2$ in water, together with the lower viscosity of acetone. The metal-organic molecules are left behind on the pore walls of the membrane after the quick evaporation of acetone upon impregnation. The thermal decomposition of the molecule converts it into metallic Pd seeds in the nano-range size. This activation method presents the advantage of producing a coating, which has adhered very well, thanks to a deeper anchorage of the Pd layer. About 50 mg of Pd-II-Acetate results in a good activation of a 10 cm long membrane.

Disadvantages in this connection include the difficultly in reproducing this activation via Pd-II-Acetate as opposed to the bath process. In order to limit the loses of Pd-II-Acetate on the container wall, the sample surface is locally impregnated, however, this makes it difficult to control the homogeneity of the dispersion. A repetition of the impregnation is advised in order to improve the distribution of the Pd-II-Acetate on the membrane surface and in the pores.

6.4. Description of the Electroless Plating process for deposition of Pd, Cu and Ag layers

6.4.1. Electroless deposition method

Electroless plating is a reliable and easily to use method for making a wide range of metal coatings on any type of support, immersed in a plating bath. The process involves the reduction of metal ions, present in the deposition bath, into their metallic form on the previously activated substrate surface. The autocatalytic deposition of the metal atoms on the seeded surface starts once the reducing agent, usually hydrazine, has been added to the plating bath. The deposition leads to the formation of a metallic layer on the substrate, which grows until one of the reactants is no longer available (ions fully reduced or reducing agent completely consumed). This method was used for obtaining different types of layers on the substrate when making composite membranes: Pd and Cu for the H_2 membrane and Ag for the O_2 membrane. The plating procedures used are based on data from references [31-33] and in-house experiments [39].

The composition of the plating bath for each coating type is given in Table 6.4-1.

Table 6.4-1: Coating bath composition for the deposition of Pd, Cu and Ag via electroless plating.

Pd coating bath	Ag coating bath	Cu coating bath
4 g $PdCl_2$ / L solution	4 g $AgNO_3$ / L solution	30 g $CuSO_4.5H_2O$ / L solution
40 g Na_2-EDTA	35 g Na_2-EDTA	60 g Na_2-EDTA
200 mL NH_3-solution (28%)	200 mL NH_3-solution (28%)	20 g NaOH
800 mL distilled H_2O	800 mL distilled H_2O	1 L distilled H_2O
Reducing agent: N_2H_4 (0.2 M)	Reducing agent: N_2H_4 (0.2 M)	Reducing agent: HCHO (37%)

Pd^{2+} and Ag^+ ions were reduced by hydrazine to their metal form; formaldehyde was used for the reduction of Cu^{2+} ions. The 3 reactions involved in the layer formation are described in Eq. 6.4-1 to 6.4-3:

$$2\ Pd(NH_3)_4^{2+} + N_2H_4 + 4\ OH^- \rightarrow 2\ Pd^0 + 8\ NH_3 + N_2 + 4\ H_2O \qquad \text{Eq. (6.4-1)}$$

$$4\ Ag^+ + N_2H_4 + 4\ OH^- \rightarrow 4\ Ag^0 + N_2 + 4\ H_2O \qquad \text{Eq. (6.4-2)}$$

$$Cu^{2+} + 2\ HCHO + 2\ OH^- \rightarrow Cu^0 + 2\ HCOOH + H_2 \qquad \text{Eq. (6.4-3)}$$

Palladium chloride has a very low solubility in distilled water, most of it remains in crystal form even where the mixture is stirred and heated. The use of an ammonia solution allows for the formation of a Pd salt, which is soluble in water due to the formation of $Pd(NH_3)_4^{2+}$ ions. These ions were then reduced by hydrazine to initiate the deposition of metallic Pd.

Na_2-EDTA was employed as a stabilizer during the deposition, as it slows down the deposition and prevents atom deposition from taking place too quickly or in an irregular manner as this would lead to the formation of porous layers. The formation of compact layers is required in order to obtain the expected gas selectivity of the composite membranes.

The steps involved in the Pd layer growth via electroless deposition are presented in the following diagrams. They do not state the exact growth mode of Pd during the electroless plating process (growth directions, grain size and grain orientation), rather supply a simplified model to illustrate coating formation, presented here as the superposition of atomic Pd layers on the Pd seeded surface. For more details on the growth mode of Pd during the electroless plating process, one can refer to Zhang et al. work, who analyzed Pd surfaces via XRD [79]. They observed that the conventional electroless deposition of Pd, as described in this chapter, leads to the formation of polycrystalline grains in 5 different orientations in the layer, namely (111), (200), (220), (311) and (222), with (111) and (311) being dominant in the coating.

1st step: Surface activation prior to deposition by anchoring Pd seeds

2nd step: Start of deposition via autocatalytic reduction of Pd^{2+} ions with hydrazine addition in the deposition bath: growth of Pd atom agglomerates at the seed location on the surface

After a certain deposition period, the deposited Pd atoms have formed a film on the support, covering the whole surface. The film thickness is determined by the amount of Pd deposited and can be increased by repeating the deposition process, when all the Pd^{2+} ions present in the bath have been reduced.

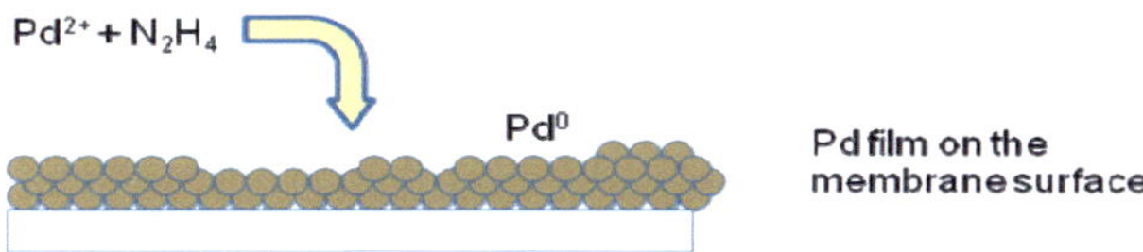

6.4.2. Experimental procedure

Palladium, copper and silver coatings were produced on activated TRUMEM® membranes. In the process, the pores of the support were progressively covered via the deposition of metallic particles; this led to the formation of a dense gas selective layer for either hydrogen (PdCu layer) or oxygen (Ag layer).

In order to realize the hydrogen selective membrane, the TiO_2 surface of the TRUMEM® support was activated via the $SnCl_2/PdCl_2$ method. After repeated activation cycles, the Pd seeds covered the whole 48 cm^2 of the membrane surface and can be recognized by their dark color. The activated support was then placed into a container of similar size, with the activated surface facing upwards. Fig. 6.4-1 shows the chemicals and material used in the electroless plating process for palladium deposition. The Pd covered membrane exhibits a metallic appearance at the end of the process. The surface morphology of the support, quite rough, is responsible for the dark metallic aspect of the Pd film, which would appear shiny on a smooth surface.

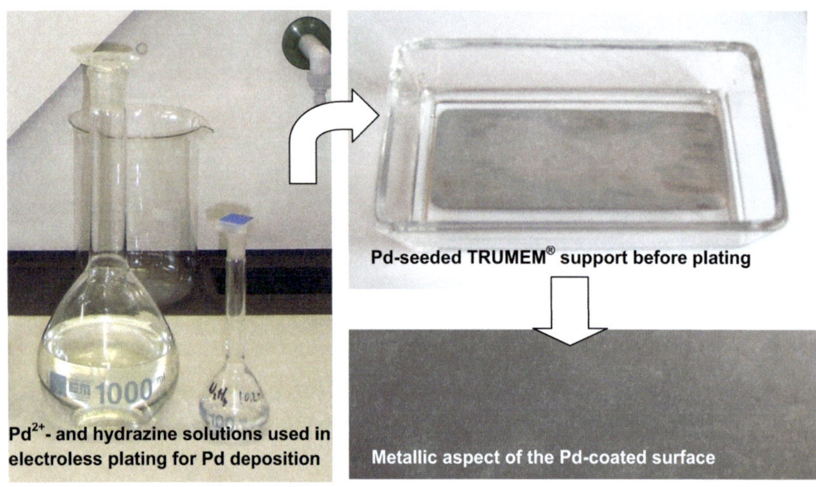

Fig. 6.4-1: View of the support and solutions used for the electroless deposition of palladium to form a hydrogen selective membrane

The plating procedure involved the following steps for all depositions.

The Pd-activated support was placed in a glass container of similar dimensions, facing upwards. The weight of the membrane was measured prior to the coating process. The container was filled with 80 mL of the plating solution.

The deposition started as soon as the reducing agent (hydrazine in the case of Pd and Ag and formaldehyde in the case of Cu) was poured into the plating solution. 1 mL was introduced first, followed by successive additions of 1 mL every 20 minutes, in order to slowly reduce the ions present in the solution. The growth of the coating was accompanied by the release of nitrogen bubbles from the surface of the membrane, in the case of Pd and Ag plating, and of hydrogen bubbles in the case of Cu deposition. When the ions from the solution were fully reduced, a gas release was no longer noticeable when another salve of reducing agent was poured into the solution. In the case of palladium deposition, 5 additions were required to reduce all the Pd^{2+} ions in the plating solution. The deposition finished once all the ions were reduced; the membrane was then taken out of the bath and rinsed several times with distilled water. The membrane was then dried at 120°C for 4 hours and weighed to estimate the thickness of the deposited film. The process was repeated until the desired layer thickness was reached. The membrane did not require any new surface activation between the 2 deposition processes as the surface was already covered by palladium.

The question of the layer thickness to be produced also had to be addressed. A too thin layer, below 5 µm, would probably not be sufficient to close all the pores present on the surface of the TRUMEM$^{®}$ support where the pores have an average diameter of 0.2 µm. This would result in defects and lead to a poor permselectivity of the membrane. On the other hand, a too thick layer, over 10 or 15 µm, would lead to a lower hydrogen permeance through the membrane and thus restrict the hydrogen partial pressure, which could be achieved in the reaction area. A good compromise is represented by a Pd-based layer thickness between 5 and 10 µm as this should allow for a satisfactory H_2/N_2 permselectivity (500 and above) and a high hydrogen permeance (in the range of 1×10^{-3} mol/m^2/s/Pa$^{0.5}$ for a temperature close to 200°C) [44].

The PdCu composite membrane was the result of the successive deposition of Pd and Cu layers, followed by a final heat treatment in N_2 at 500°C for 24h. A Pd layer could not be produced on top of a Cu layer without prior heat treatment, as the Pd solution would have hydrolyzed the Cu layer if it had not been previously alloyed.

In the 5 µm PdCu membrane made by electroless plating, a Pd content of 59 wt.% was determined based on the weights of the two layers deposited. This $Pd_{59}Cu_{41}$ wt.% membrane was used in experiments described further on in this document.

Fig. 6.4-2 is a view of a 5 µm Ag membrane, produced by electroless plating, on a TRUMEM$^{®}$ support and positioned in the reactor half. Similar to the PdCu composite membrane, the 5 µm Ag layer was heat treated at 500°C for 24h in order to homogenize the successively deposited layers (repetition of the process until 5 µm is reached). The white appearance of the surface of the Ag layer contrasts with the metallic appearance of the Pd layer, which was produced via the same method (see Fig. 6.4-1). The two types of gas selective membranes were characterized to determine their respective hydrogen (through the PdCu layer) and oxygen (through Ag) transport properties.

Fig. 6.4-2: View of the surface of the 5 µm Ag plated membrane placed in the reactor

Fig. 6.4-3 illustrates the surface morphology of the Pd, Cu and Ag layers, which were produced by electroless plating. A difference in the surface morphology becomes apparent. Finer agglomerated particles are observed in the case of Pd and Ag. The growth of the Pd atoms to form crystallites (compared to the round Ag particles) should result in a high surface area for the hydroxylation reaction, as Pd acts simultaneously as a hydrogen diffusion layer and catalyst for benzene hydroxylation.

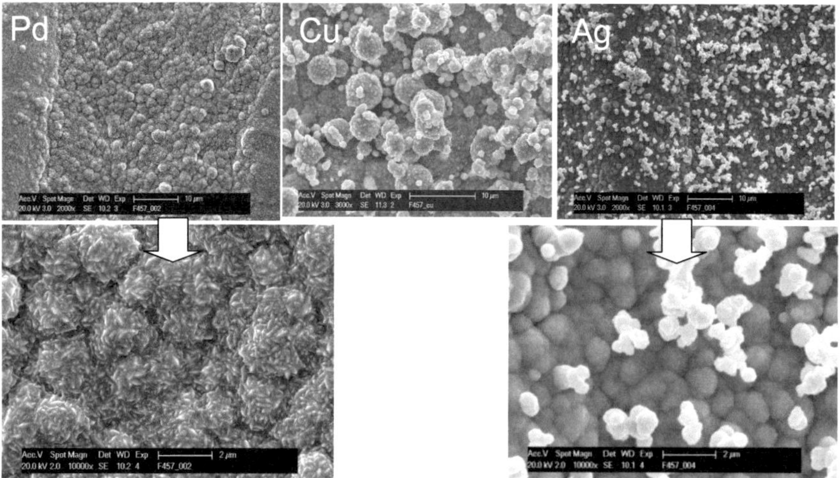

Fig. 6.4-3: SEM images of the surface morphology of Pd, Cu and an Ag coatings produced by electroless plating on TRUMEM® supports before heat treatment

The surface morphology of the Pd layer as it appears in Fig. 6.4-3 with crystallites oriented in different "random" directions seems to correlate the observations of Zhang et al. on the growth of Pd grains in 5 directions during the deposition process [79].

6.5. Modification of the catalytic properties of the $Pd_{60}Cu_{40}$ membrane

6.5.1. Purpose

In order to improve the phenol rate and selectivity of the initial PdCu membrane, a modification of the surface state was realized by means of the deposition of more active catalysts for benzene hydroxylation. Three different catalyst types were considered and tested, with the aim of reducing the influence of side-reactions as is described in part 9. The surface modification of the foil was based on the following systems.

6.5.2. Palladium-gold

OH-species are believed to be essential to convert gas phase benzene into phenol in a one-step reaction. The amount and life time of these radical species on the membrane will have an influence on the observed phenol rates. A method for increasing the phenol rate could be the use of a Pd-based catalyst with improved activity vis-à-vis the formation of OH-radicals. Investigating the liquid-phase formation of hydrogen peroxide over different catalyst types, Edwards et. al obtained higher productivities with PdAu catalysts [49-51]. The application of a thin PdAu layer on top of the PdCu membrane could generate more OH-radicals, which are able to hydroxylate benzene into phenol. This option was tested.

The PdAu alloy is present, nearly over the whole composition range, as a solid solution. It forms Pd_3Au and $PdAu_3$ intermetallic phases only at around 1000 K (see Fig. 6.5-1) and PdAu at room temperature close to a composition of 50 wt.% Pd. Gold and palladium have atomic radii of 1.35 Å and 1.37 Å, respectively, and can easily substitute each other in their respective FCC lattices.

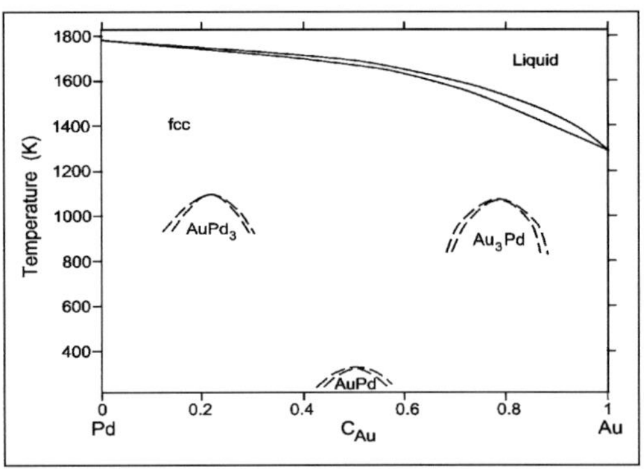

Fig. 6.5-1: Pd-Au phase diagram

The combination of Pd and Au (up to 50 wt.% Au) has been reported to be very active towards the H_2O_2 synthesis [49-51]. Edwards et. al obtained a successful hydrogen peroxide synthesis on a TiO_2-supported Au-Pd catalyst, with H_2O_2 selectivities of 93% achieved under appropriated conditions and with pre-treatment [49-51]. Palladium and gold are fully miscible over the whole composition range. I opted for a low gold content $Pd_{90}Au_{10}$ (wt.%) catalyst, due to its good hydrogen permeability, which was manufactured as target material and was deposited on the flat $Pd_{60}Cu_{40}$ membrane surface via magnetron sputtering.

6.5.3. Palladium-gallium

A PhD. thesis from TU Berlin, 2005, aimed at investigating the catalytic hydrogenation of acetylene using *PdSn* and PdGa catalysts presented a concept, known as active-site isolation, which depicts the influence of a particular Pd surface composition on the adsorption/desorption behavior of gas phase molecules in the case a hydrogenation reaction [75-76]. An interesting intermetallic Pd-based catalyst was described. A $Pd_{50}Ga_{50}$ (at. %) catalyst of this exact composition possesses the particularity of having each of its palladium atoms strictly neighbored by a gallium atom, which creates a surface pattern where all the active Pd sites are isolated from each by "spacing" Ga atoms (see Pd-Ga phase diagram of lattice structure in Fig. 6.5-2).

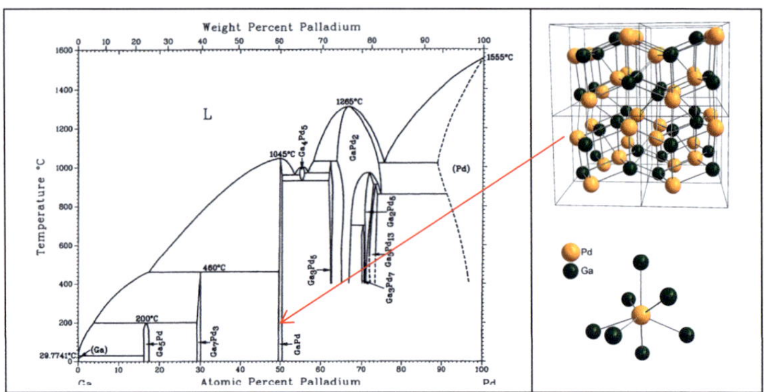

Fig.6.5-2: Pd-Ga phase diagram and lattice structure of $Pd_{50}Ga_{50}$ (at. %)

As shown in Fig. 6.5-2, $Pd_{50}Ga_{50}$ at. % has a well ordered crystal structure with Pd atoms solely coordinated by Ga atoms. This structure disables any strong adsorption of gas phase molecules due to the absence of 3 adjacent Pd sites. A weaker "α-adsorption" of the molecule on the Pd surface, therefore, leads to quicker adsorption/desorption times in the gas phase. In the case of a benzene molecule adsorbed on this catalyst, a stronger adsorption might lead to an increased total oxidation. A weaker bond of the molecule could thus increase the phenol selectivity by limiting the side-reactions. Ga atoms should be inert; they are not known to have catalytic properties in a hydrogen-related reaction system. The geometric configuration with pure Pd and $Pd_{50}Ga_{50}$ (at. %) is presented in Fig. 6.5-3.

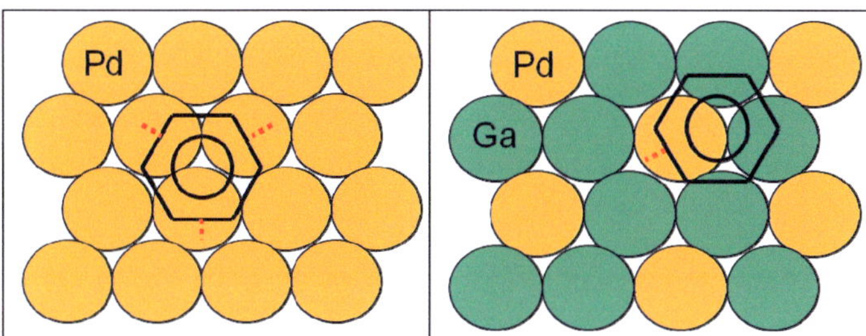

Fig. 6.5-3: Expected adsorption-desorption behavior of benzene on a Pd and Pd$_{50}$Ga$_{50}$ (at. %) surface. A quicker adsorption/desorption kinetic of benzene is expected on isolated Pd atoms.

As this adsorption phenomenon has been observed in the case of the selective hydrogenation of acetylene (C-C distance in acetylene of about 1.20 Å) [75-76], a similar behavior can be expected with gas phase benzene molecules, where the distance between 2 carbon atoms is about 1.40 Å. As the Pd-Pd distance in the fcc lattice is 2.75 Å, spacing Ga atoms would disable a stronger adsorption of benzene on neighboring Pd atoms.

6.5.4. Ceramic particle-supported metal catalyst system: V$_2$O$_5$/PdAu

The idea behind this system is to increase the catalyst surface area by depositing ceramic particles with an insular morphology (rough surface); the catalyst is, in turn, deposited on top of this. Oxide particles are initially deposited on the PdCu foil as a support for the Pd-based particles, which are applied in a second step. Vanadium oxide has been reported to be efficient in the case of benzene to phenol in the liquid phase [65-66]. PdAu was chosen for the same reason as for why it was used as a compact layer, namely for its higher activity towards hydrogen peroxide. V$_2$O$_5$ particles are produced by the brief sputtering of vanadium under reactive atmosphere in the presence of oxygen; PdAu particles are similarly produced on the surface by a short sputtering of PdAu in Ar atmosphere (from seconds to minutes).

Kunai et al. reported that phenol was obtained with almost 100% selectivity on a Cu-Pd/SiO$_2$ catalyst by the gas phase hydroxylation of benzene in the presence of O$_2$ and H$_2$ [17]. Kitano et al. noticed that no catalytic effect appears, if one of the metal species is missing, where a Pd-Cu SiO$_2$ supported catalyst is used [20].

Fig. 6.5-4 summarizes the 3 catalytic configurations to be realized on the Pd-based membranes. The first two are based on layers, while the third is based on supported particles.

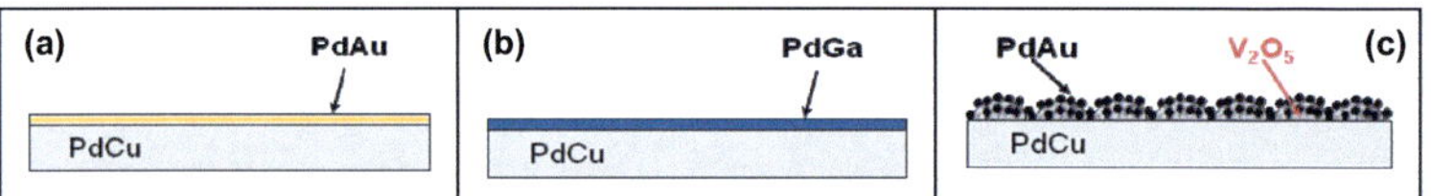

Fig. 6.5-4: Modification of the PdCu foils by the magnetron sputtering of different catalytic systems: chemical (a), sorption (b) and surface area (c) effects are expected.

6.6. Catalyst deposition by means of magnetron sputtering

6.6.1. Principle and composition

Magnetron sputtering is a physical deposition process, which enables the deposition of metallic and ceramic materials, of simple or complex composition, on any type of support without requiring any kind of pretreatment (apart from cleaning). It features similar advantages to a chemical vapor deposition process; however, a sputtering deposition is most suited for coating flat surfaces, as the gas phase atoms to be deposited cannot be "guided" as is the case in the chemical vapor process. The "physical" adhesion of the layer on the sample to be coated is guaranteed via the location of the sample on the system anode (bottom part), whereas the ablative target disk is placed at the cathode (upper part). The deposition takes place in a so-called deposition chamber where a process gas is injected (usually argon). By applying a voltage between the anode and cathode, the process gas is ionized and creates a plasma of Ar^+ ions and electrons. While the electrons are accelerated to the anode, the positive ions are accelerated to the cathode on which the ablative target is fixed. When they collide with the target, the impact energy expels atoms from the target surface; these "fall" onto the bottom of the chamber (anode) and continue to form a layer until the end of the deposition process. The principle of magnetron sputtering is illustrated in Fig. 6.6-1.

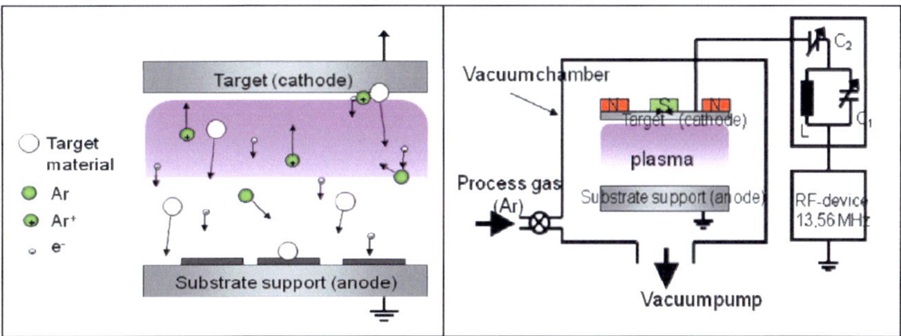

Fig. 6.6-1: Principle of substrate coating by magnetron sputtering.

The intensity of the plasma is controlled by the voltage applied between anode and cathode, which can be produced either by a direct current source or by radio frequencies. The possibility exists to use a reactive gas in addition to argon in order to form an oxide or a nitride layer using O_2 or N_2 respectively. The "magnetron" term means that magnets are coupled to the cathode. The magnetic field improves the efficiency of the deposition, as it orients the plasma towards a larger area of the target. The ablation of the target occurs by forming a ring, which limits the peripheral losses.

Fig. 6.6-2 shows the magnetron sputtering device available for this work at the Karl-Winnacker-Institute. The deposition chamber, cathodes and a sputtering target are shown in detail.

Fig. 6.6-2: Pictures of the Magnetron Sputtering Device of the Karl-Winnacker-Institute.

The magnetron sputtering machine is equipped with 4 cathode sources, enabling the co-deposition, where required, of 4 elements at the same time. Two sources are controlled by radio frequencies, the others by direct current. Depositions via direct current are preferred in this work for practical reasons as the deposition rate of the target material is directly proportional to the applied DC voltage and is, in general, higher than for RF sputtering (sinusoidal signal controlling the plasma acceleration).

The process parameters influencing the deposition kinetics are the following:

- nature of target
- sample – target distance (or anode – cathode)
- voltage source (DC or RF) and amplitude
- deposition time
- vacuum level in the chamber
- presence or absence of a reactive gas

In order to coat the PdCu foils with a catalytic layer or a pattern of catalytic particles, they were placed on the anode of the system and sputtered with the desired elements, which were manufactured in the target shape by the company Lesker Ltd (USA).

6.6.2. Deposition parameters and analysis of catalytic layers

Preliminary deposition experiments have allowed for the determination of the deposition rate of the desired elements using standard parameters of the magnetron sputtering device.

Table 6.6-1: Sputtering parameters for deposition of $Pd_{90}Au_{10}$, $V_2O_5/Pd_{90}Au_{10}$ and $Pd_{50}Ga_{50}$

Target source	$Pd_{90}Au_{10}$	$Pd_{50}Ga_{50}$	$Pd_{50}Ga_{50}$	$Pd_{50}Ga_{50}$	V
Layer / particles (l/p) to be deposited	$Pd_{90}Au_{10}$ (l/p)	$Pd_{50}Ga_{50}$ (layer 5 µm)	$Pd_{50}Ga_{50}$ (layer 1.4 µm)	$Pd_{50}Ga_{50}$ (layer 0.4 µm)	V_2O_5 (p) via reactive sputtering
Atmosphere composition in deposition chamber	100% Ar	100% Ar	100% Ar	100% Ar	60% Ar 40% O_2
Pressure in the chamber (mbar)	1.10^{-3}	1.10^{-3}	1.10^{-3}	1.10^{-3}	1.10^{-3}
DC voltage (kV)	0.1	0.1	0.07	0.07	0.1
Target-substrate distance	standard (10 cm)	standard	standard	standard	standard
Experimental growth rate (nm/min)	70	90	90	90	20
Sputtering time	1 min (p) to 15 min (l)	60 min (l)	15 min (l)	5 min (l)	1 min (p)

The 3 catalytic-surface modification types have been realized according to the data indicated in Table 6.6-1. Three $Pd_{60}Cu_{40}$ foils, 50 µm thick, have been successively modified by the magnetron sputtering of a $Pd_{90}Au_{10}$ layer, a

$Pd_{50}Ga_{50}$ layer and $V_2O_5/Pd_{90}Au_{10}$ particles, for the purpose of obtaining the configurations sketched in Fig. 6.5-4.

After deposition, samples of the produced coatings were analyzed by means of scanning electron microscopy and EDX (model Philips XL40 equipped with LaB_6 electron gun) in order to identify the morphology of the catalysts and determine their composition after deposition on the support. The composition should ideally be identical with that of the initial sputter target. It is, however, possible that the target materials are deposited as separated elements with different deposition rates instead of as homogeneous alloys. In this case, the composition of the deposited layer would differ to some extent from that of the sputter source.

Analysis of the $Pd_{90}Au_{10}$ (wt. %) layer:

Fig. 6.6-3 shows the surface of the PdAu layer deposited from a $Pd_{90}Au_{10}$ wt.% sputtering source, as well as the EDX analysis of two areas of this surface.

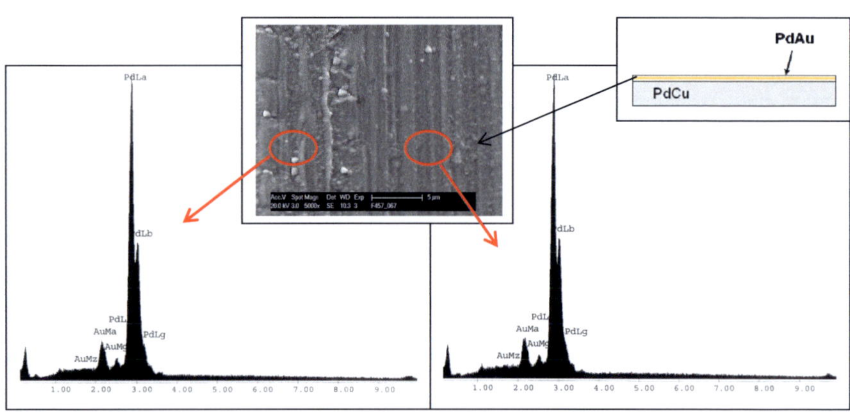

Element	Wt. %	At. %
Au M	9.76	5.52
Pd K	90.24	94.48
Total	100.00	100.00

Element	Wt. %	At. %
Au M	10.46	5.94
Pd K	89.54	94.06
Total	100.00	100.00

Fig. 6.6-3: SEM surface image and EDX spectra of a PdAu layer sputtered from a $Pd_{90}Au_{10}$ wt.% target.

From the SEM surface image, the main observation is the presence of many cold rolling traces, which are due to the manufacturing process of the PdCu foils on which the layer has been formed. The PdAu layer logically reproduces the same topology. More interesting is the result of the EDX analysis. Only Pd and Au are detected, no trace of contaminant or support element was identified. The Pd wt.% analysis indicated that the areas tested have a composition, which varies slightly from that that of the initial sputter target, with 90.24% and 89.54%, respectively, obtained (90.00% was expected). However, the scatter values are inside the error range for the measurements (0.41% – 0.46%). It can, therefore, be considered that the average value of 90.0% wt.% Pd corresponds with the amount of Pd effectively deposited on the PdCu support. In this case, a homogeneous deposition of the PdAu catalyst was observed.

Relevant supplementary XRD analyses were carried out, using a STOE-STADIP-MP powder diffractometer with a Cu-Ka1 ray of wavelength 1.5406 Å, at MPI Dresden in order to identify the phases present in the new catalytic surfaces.

Fig. 6.6-4 shows the XRD analysis of the surface with the 1 µm PdAu-sputtered PdCu foil. Only one type of reflex was detected, which corresponds to the PdAu phase where Pd and Au are in solid solution in the alloy. The measured PdAu reflex deviates slightly from its expected position; this is due to the angle at which the sample surface was impacted by the X-rays during the respective measurements. The PdCu foil below the layer was not detected during the sample scanning.

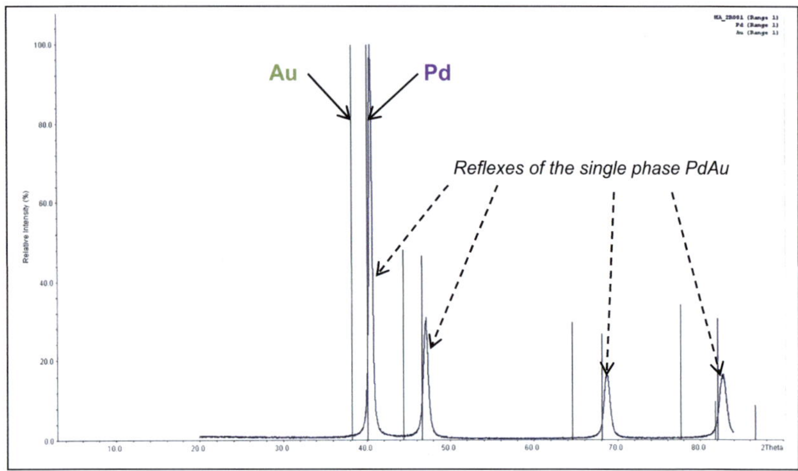

Fig. 6.6-4: XRD spectrum of a PdAu layer sputtered from a $Pd_{90}Au_{10}$ wt.% target.

The XRD analysis confirms that none of the alloy elements were deposited as single metals during the sputtering process. A layer with the same composition as the sputtering target was produced.

A cross-section of a surface-modified membrane was also investigated via ESMA analysis using a JXA-85360F Microsonde; this was carried out at the IMVT-KIT in Karlsruhe (post experimental analysis carried out after completion of the laboratory work). The purpose was to detect possible element concentration gradients within the layer and identify whether Cu could diffuse to the surface during the course of the hydroxylation experiments and thus modify the catalyst properties.

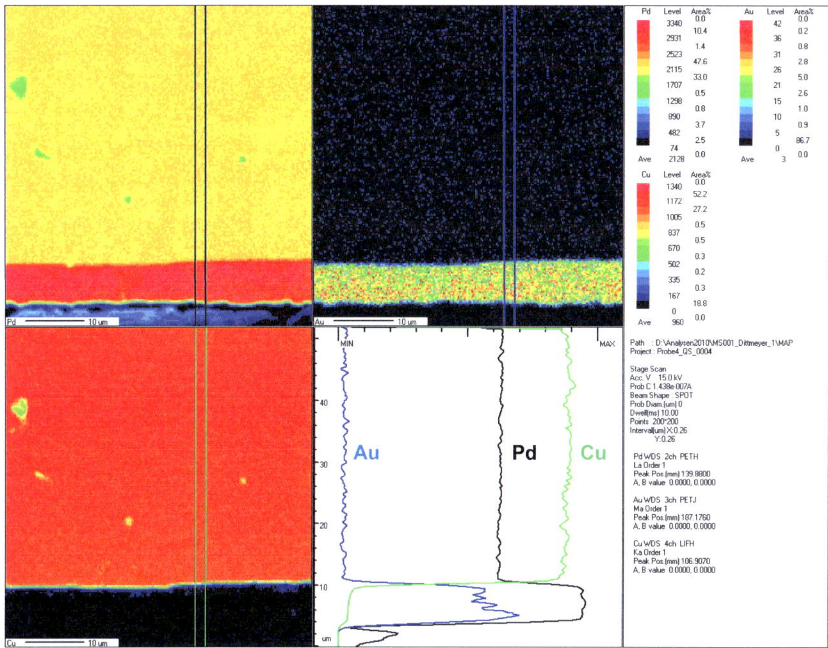

Fig. 6.6-5: ESMA elemental line-scan of a PdAu layer sputtered over a PdCu foil from a $Pd_{90}Au_{10}$ wt.% target.

It was initially observed that the PdAu layer, which was extracted as a sample from one end of the membrane, unexpectedly presented a thickness, which was greater than the average 1 µm calculated after catalyst sputtering (around 5 µm). This indicates that the sputtering process may lead to irregular thicknesses along the membrane length (the sputtering source was located close to one end of the membrane). A higher Pd concentration, which may be due to bulk diffusion after repeated hydroxylation experiments at 150°C, was also observed in a zone about 6 µm thick below the coating. A low concentration gradient of Cu in the PdAu coating, which is also due to bulk diffusion effects at the same temperature, was also detected. The concentration of Cu, however, at the catalyst surface appears to be null; a lowering of the PdAu catalyst efficiency, due to contamination by a third element, is thus not expected.

Analysis of the $Pd_{50}Ga_{50}$ (wt. %) layer:

The surface of the PdGa layer deposited from a $Pd_{50}Ga_{50}$ wt.% sputtering source and its EDX surface analysis are presented in Fig. 6.6-6.

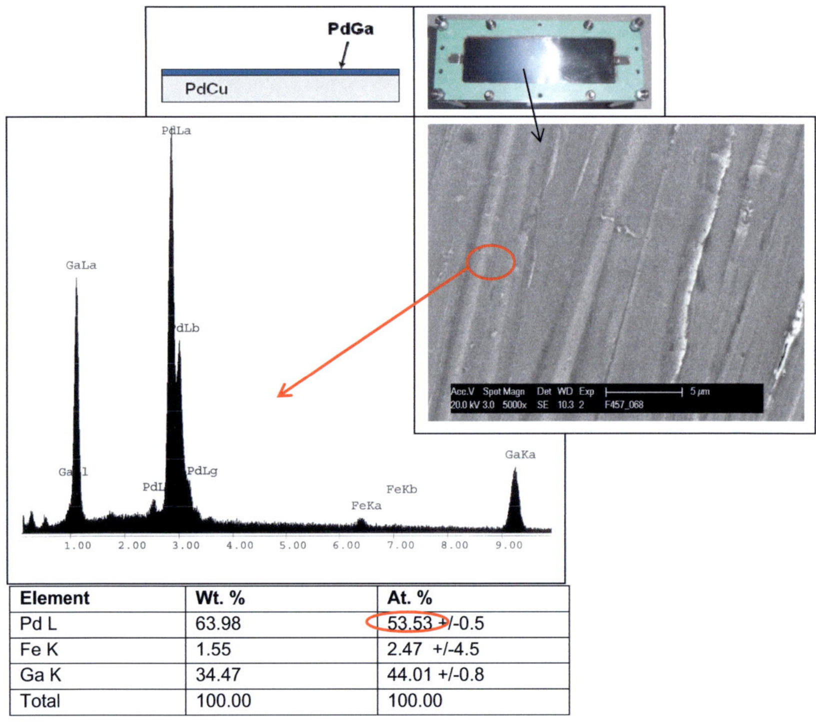

Element	Wt. %	At. %
Pd L	63.98	53.53 +/-0.5
Fe K	1.55	2.47 +/-4.5
Ga K	34.47	44.01 +/-0.8
Total	100.00	100.00

Fig. 6.6-6: SEM surface image and EDX spectrum of a PdGa layer sputtered from a $Pd_{50}Ga_{50}$ at.% target.

The cold rolling traces from the support underneath was observed with respect to the surface morphology of the PdGa coated foil. No particle agglomeration or inhomogeneous deposition was noticed. The EDX spectrum of the sample surface shows traces of iron, an impurity, this is, however, within the error range. These traces probably come from the area where the foil was in contact with the reactor wall. The atomic analysis indicated that a higher amount of Pd was deposited in comparison to gallium, whereas an equal deposition of 50 at.% was expected for both elements. When balancing the amounts of Pd and Ga, excluding the iron impurities, one obtains 55 at.% of Pd and 45 at.% of Ga (the measurement error for Pd and Ga is below 1 at.%). The deposited catalyst does not have the exact same composition as the sputter target from which it is produced. This composition shift will affect the properties of the catalytic layer, as pure intermetallic phase $Pd_{50}Ga_{50}$ is the target catalyst, which should be produced on the PdCu foil in order to improve the phenol selectivity. Because the composition range is close to 50 at.% Pd, a large area of the catalyst surface might, however, be covered by the intermetallic phase. An XRD analysis of the surface is required in order to investigate the nature of the phases co-deposited with $Pd_{50}Ga_{50}$ during sputtering. Based on the phase diagram of Pd-Ga (see Fig. 6.5-2), several phases exhibiting an excess of Pd could be present in the layer: Pd_5Ga_3, Pd_2Ga or even pure Pd grains. Fig. 6.6-7 shows the results of the XRD analysis performed on samples of with a thickness of 5 µm, 1.4 µm and 0.4 µm.

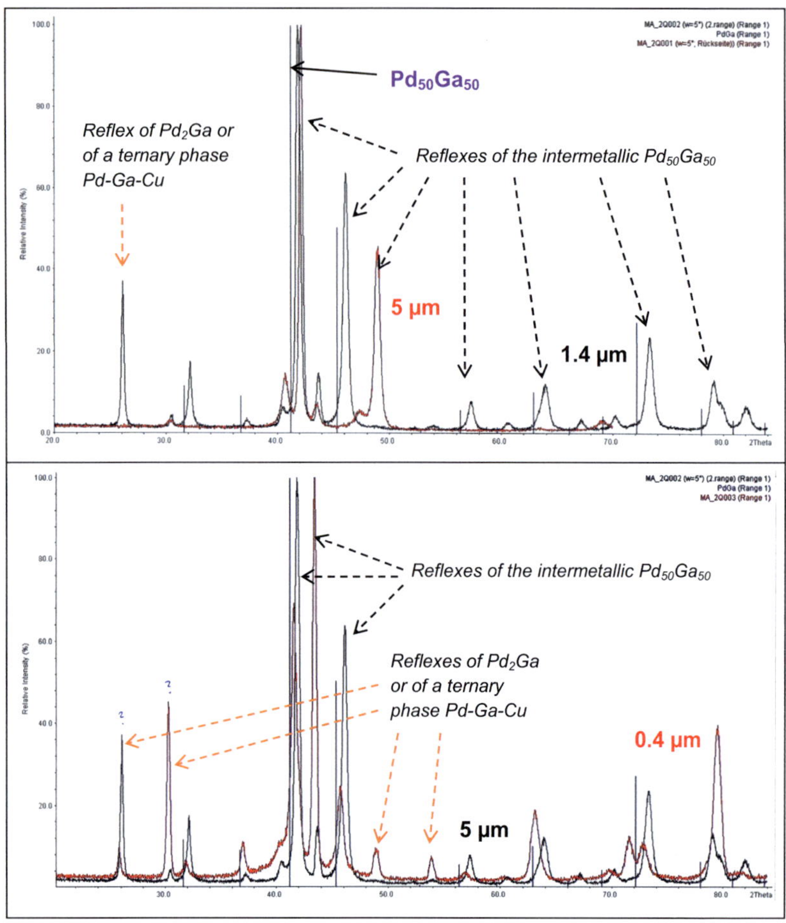

Fig. 6.6-7: XRD spectra of 5 µm, 1.4 µm and 0.4 µm PdGa layers sputtered from a $Pd_{50}Ga_{50}$ at.% target.

The XRD spectra reveal the presence of 2 phases. The main one, $Pd_{50}Ga_{50}$, is identified without any doubt as the reflexes appear at all the expected 2Theta angles. The presence of another phase is also shown, with reflexes at 2Theta angles equal to 26°, 30°, 49° and 54°. Reflexes at these angles cannot be attributed with certainty to an identified phase, however, 2 options emerge. They do not correspond to pure Pd, but they could be from other Pd-rich phases, like Pd_2Ga or a ternary alloy based on Pd-Ga-Cu. Due to the inertness of intermetallic PdGa towards the formation of hydride, the hydrogen transport through the layer formed could be controlled by diffusion through this Pd-enriched phase.

Fig 6.6-8 shows the result of the ESMA analysis performed on the cross section of a 5 µm thick PdGa sputtered foil (post-experimental analysis).

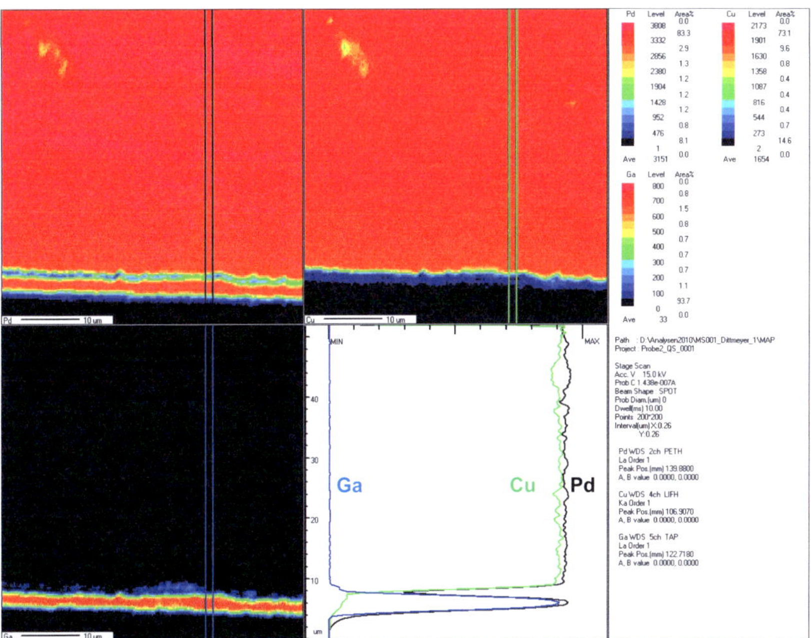

Fig. 6.6-8: ESMA elemental line-scan of a PdAu layer sputtered over a PdCu foil from a $Pd_{50}Ga_{50}$ at.% target.

The sputtered PdGa layer is detached from its support; this was probably caused by shear stresses, which occurred during the sample cutting and preparation. The thickness of the examined PdGa layer (ca. 4 µm) is close to that calculated after catalyst deposition (5 µm). There is, in accordance with the line-scan, also a slight diffusion of Cu from the PdCu foil into the PdGa layer, as was previously observed with the PdAu film. The presence of a Cu concentration gradient in the PdGa layer confirms the possible formation of a ternary Pd-Ga-Cu phase, as was detected by means of X-ray diffraction.

The same is observed for samples extracted from the 0.4 µm and 1.4 µm thick PdGa films. These results are therefore not displayed.

Analysis of the surface sputtered with $V_2O_5/Pd_{90}Au_{10}$ particles

The V_2O_5/PdAu-sputtered catalyst was analyzed using EDX both at its surface and sub-surface at the interface level with the support in order to establish its composition after the sputtering process. Fig. 6.6-9 shows the topology of both faces in combination with the initial concept (increase of the catalyst surface area via particle deposition) and shows the result of the EDX compositional analysis.

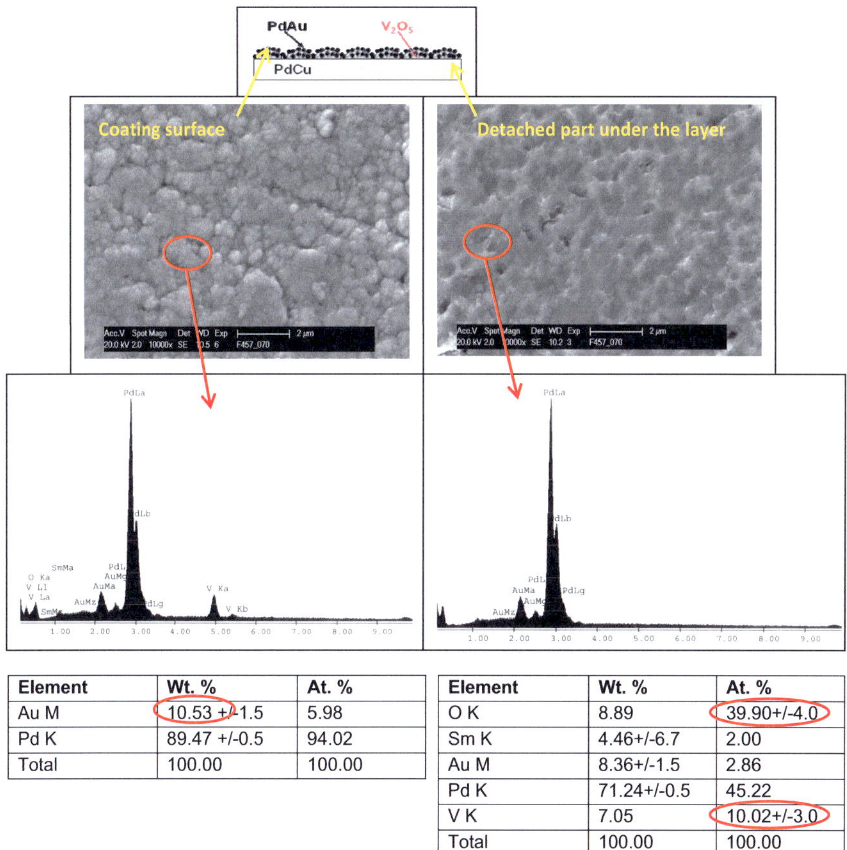

Element	Wt. %	At. %
Au M	10.53 +/-1.5	5.98
Pd K	89.47 +/-0.5	94.02
Total	100.00	100.00

Element	Wt. %	At. %
O K	8.89	39.90+/-4.0
Sm K	4.46+/-6.7	2.00
Au M	8.36+/-1.5	2.86
Pd K	71.24+/-0.5	45.22
V K	7.05	10.02+/-3.0
Total	100.00	100.00

Fig. 6.6-9: SEM surface image and EDX spectrum of a V_2O_5/PdAu sputtered surface using a 100% V and a $Pd_{90}Au_{10}$ wt.% targets.

Visually one can see that the surface topography obtained after the sputtering of the 2 elements more or less corresponds to that expected, i.e. an insular morphology based on oxide particles, which permit hydrogen diffusion to the catalytic particles on top of it. A higher surface area was expected due to the PdAu particles deposited on the insular support, however, the elemental surface analysis shows a deviation from this objective. Only PdAu is, in fact, found on the surface in spite of the short sputtering period; this indicates that the brief sputtering process produced a layer of at least 100 nm (depth of EDX emission) and not just PdAu particles on the V_2O_5 support. The effect thus produced by this catalyst will probably not fulfill the initial expectation due to the limited reaction surface area provided by a layer in comparison to that which would be provided by a surface area sputtered with particles. The sputtered elements present underneath the PdAu layer were also investigated by peeling off a piece of catalyst from the PdCu foil and analyzing said using EDX. The analysis revealed the presence of V and O in a total at.% amount of 49.9%, which tends to indicate that about half of the foil surface was covered by vanadium oxide after a 1 minute sputtering process. Pd, Au and traces of contaminant (Sm) were detected in addition to V and O. The error margin for the Sm measurement is higher than the amount detected; one can thus exclude it from the catalyst composition.

Fig. 6.6-10 illustrates the XRD analysis of the samples of both sides of the V_2O_5/PdAu catalyst.

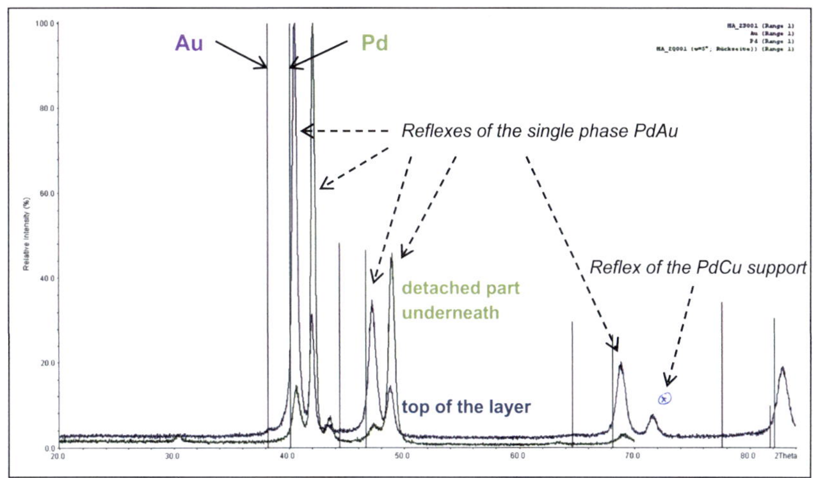

Fig. 6.6-10: XRD spectra of a V_2O_5/PdAu layer sputtered from V and $Pd_{90}Au_{10}$ wt.% targets.

The amorphous ceramic phase (V_2O_5) is not detected, as it has no crystal structure. The peaks of the PdAu phase clearly appear for both samples. A small reflex, which corresponds with the PdCu support under the catalyst is obtained in one case. This could be explained by the fact that, even though a PdAu layer was produced instead of particles, its thickness is rather low (range of 100 nm) which let the X-Rays attain the support. Some PdCu particles from the foil may potentially have diffused to the catalyst surface during the 2-step sputtering process.

Fig. 6.6-11 illustrates the ESMA analysis of the V_2O_5/PdAu catalyst deposited on PdCu. The distribution of the different elements (Pd, Cu. Au. V and O), present in the sample, was investigated and appears in the below line-scans.

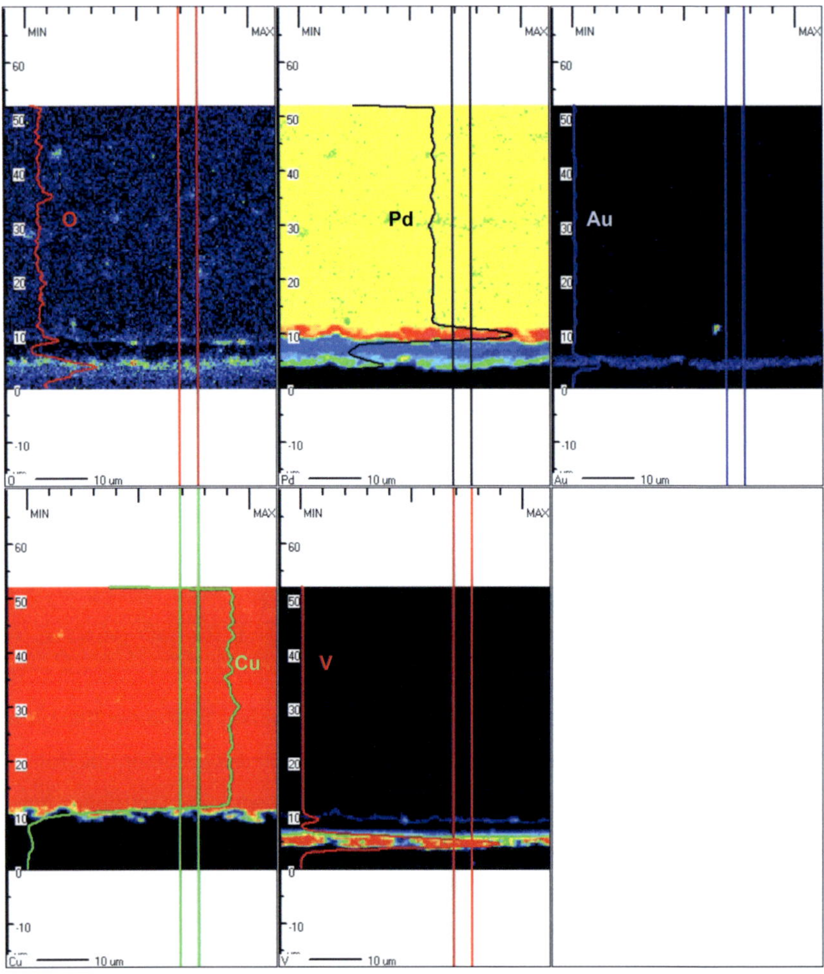

Fig. 6.6-11: ESMA elemental line-scan of a V_2O_5/PdAu layer sputtered from V and $Pd_{90}Au_{10}$ wt.% targets over a PdCu foil.

Some Pd diffusion from the foil to the surface at the interface with the vanadium oxide particles, which form an interrupted layer, on top of which the second catalyst is applied, was again noticed. It can be observed that the briefly sputtered PdAu catalyst has formed a thin film on V_2O_5 and not agglomerated particles as was initially expected. No Cu seems to have diffused to the top catalytic part, as was the case for the PdGa systems. Only a small peak of PdCu from the support was detected via XRD measurements.

6.7. Description of the sol-gel process for the preparation of porous UF membranes

The intention here is to produce UF membrane layers on the commercial support with a nominal pore size between 3 and 20 nm via a sol-gel method, in order to obtain a homogeneous oxygen distribution without using silver. A replacement of the silver membrane, with a UF membrane with dispersed pores of a well defined size, had to be considered due to the fact that the silver membrane may not be capable of delivering sufficient oxygen flows, in comparison to the hydrogen supply, through Pd-based membranes. The composite TRUMEM® membrane with its nominal pore size of 200 µm is also used as a support for the UF coating. It has the advantage of already having a TiO_2 top layer; adhesion to the next UF TiO_2 layer will thus be easier.

Experimental procedure [57-59]:

A colloidal TiO_2-sol is initially prepared via the hydrolysis of titaniumalkoxide and peptisation by diethanolamine. The sol contains 2.1 wt.% TiO_2. A sol solution is then prepared by adding a binder, in this case a 2 wt.% solution of hydroxyethyl-cellulose (HEC), and is homogenized on a rotating plate for a few hours before application. The sol solution is then spread, using a brush, on the 48 cm^2 TRUMEM® membrane surface forming a gel layer on it; the membrane was cleaned and dried beforehand. Two heat treatments are subsequently realized, a first one at 300°C for 3 hours, to dry the surface and burn off any organic constituents; then a sintering heat treatment at 500°C for 6 hours to transform the gel layer into a UF ceramic layer. The thickness of the UF layer obtained is influenced by the viscosity and amount used of the sol solution, as well as by how the layer is applied. A layer thickness of about 50 µm was experimentally produced on membrane samples using this procedure.

SEM analysis of the surface of the UF layer:

Fig. 6.7-1 shows SEM images of the surface of the TRUMEM® membrane before and after deposition of the UF TiO_2-layer. No cracks or defects appear on the surface. It is evident that the surface pore size has been reduced by the production of the UF layer. The SEM investigation of the sample lead to the same pattern of small pores well distributed over the entire surface; this tends to indicate that the sol-gel layer was homogeneously applied before sintering. The homogeneity of the pore distribution is a condition for obtaining a well-defined dosage of oxygen through the whole surface of the membrane.

Fig. 6.7-1: SEM views of the surface of the uncoated TRUMEN® surface (image on left) and of a UF-TiO₂ layer deposited on a TRUMEM support by the sol-gel method (images at center and on right)

The smallest pore size, which can be attained, depends on the size of the sol particles, which is around 20 nm; it is thus not possible to produce pores smaller than 2 nm. NF layers would have to be prepared if required.

7. Membrane characterization by means of permeation measurements

Prior to the hydroxylation experiments, all prepared membranes were characterized by gas permeation measurements in order to determine their transport properties for hydrogen. This characterization has a double purpose. Firstly to determine the process parameters, in particular the retentate partial pressure, which allow for the control of the amount of hydrogen introduced in the reactor during the benzene hydroxylation at a given temperature. Secondly to allow for a correlation between the reaction performance, which depends on the membrane catalytic properties and hydroxylation conditions, and the reached hydrogen partial pressure level for the same set of process parameters using the different membranes. A layer featuring a too low hydrogen permeability could, in fact, limit the phenol rate due to the low level of hydrogen partial pressure reached for the reaction in comparison to a membrane that would be catalytically less active but more hydrogen-permeable.

7.1. Membranes for hydrogen dosage:

The experimental procedure employed for the determination of the permeation properties of the membranes is as follows. The hydrogen permeation measurements for all Pd-based membranes were realized with a nitrogen sweep flow of 80 mL/min at the permeate side of the membrane in the central reaction channel; in this connection the temperature was varied between 150°C and 180°C, while the hydrogen partial pressure at the retentate side was varied between 1 and 3 bar. Due to the presence of the nitrogen sweep in the reactor, the permeate hydrogen partial pressure was determined using the µGC, for the analysis of the gas phase composition, and the measurement of the gas flow rate exiting the reactor at 1 bar total pressure. The total flow was measured at room temperature before entering the µGC, as illustrated in Fig. 7.1-0.

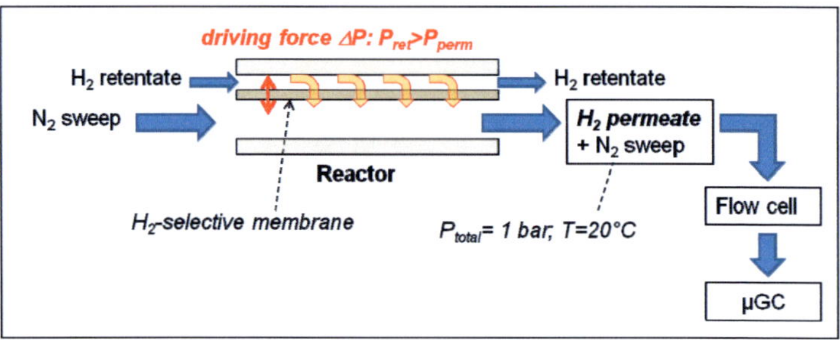

Fig. 7.1-0: Diagram of the laboratory system used for H_2 permeation measurement with the different prepared membranes

The same system was employed for the oxygen permeation measurements presented in part 7.2, whereby the H_2-selective membrane was replaced by the O_2-selective one.

7.1.1. PdCu foil

The $Pd_{60}Cu_{40}$ wt.% alloy was the standard H_2 permeable material for the dosage of active hydrogen species in the system for performing the hydroxylation reaction. A self-supported PdCu foil and a supported electroless-plated PdCu layer were tested in order to determine the membrane type with the best performance; this can be influenced by the preparation process independently of the reaction parameters. The PdCu foil was initially characterized. It was produced by cold-rolling down to a nominal thickness of 50 µm, the lower limit allowed by the process (MaTeck, Germany). The foil was not submitted to any heat treatment before the permeation measurements.

Fig. 7.1-1 illustrates the hydrogen flux measured through the 50 µm thick PdCu foil for different temperatures and the hydrogen partial pressure differences applied at the retentate side. The hydrogen flux is plotted versus the hydrogen partial pressure difference at the exponent n=0.5, as suggested by Sieverts's law usually applying to hydrogen permeation through Pd-based membranes.

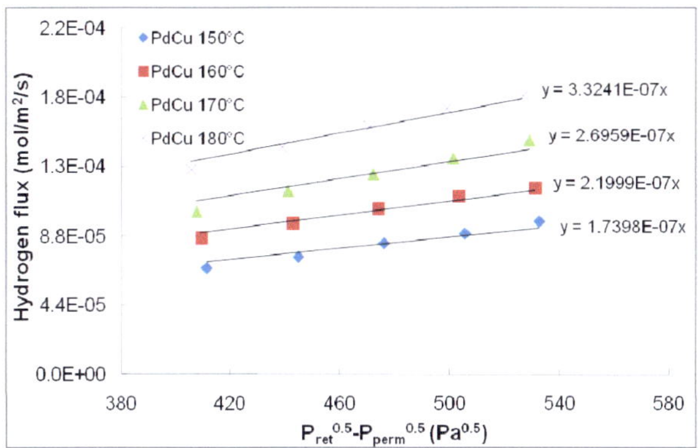

Fig. 7.1-1: H_2 permeation measurements carried out on a 50 µm thick $Pd_{60}Cu_{40}$ foil. Influence of the temperature and partial pressure difference on the permeating flux.

The diffusion of hydrogen through the bulk of the membrane is thermally activated, as illustrated by the flux increase together with the temperature at constant driving force.

As expected for a rather thick Pd alloy foil, the hydrogen flux data in Fig. 7.1-1 can be well reproduced using Sieverts's law:

$$J_{H2} = Q.S.\left(P_{H2\,ret}^{\;n} - P_{H2\,perm}^{\;n}\right) \qquad \text{Eq. (7.1-1)}$$

with:

J_{H2}: hydrogen flux through the membrane (mol.m^{-2}.s^{-1});

Q: membrane permeance (mol^{-1}.m^{-2}.s^{-1}.Pa^{-n});

P_{H2ret}: hydrogen partial pressure at the retentate side (Pa);

P_{H2perm}: hydrogen partial pressure at the permeate side (Pa);

n: Sieverts's exponent, equal to 0.5 for a pure Pd layer ranging from approx. 1 µm to 100 µm in thickness. In this case, bulk diffusion of the dissociated hydrogen species controls the permeation behavior through the membrane. n-values ranging between 0.5 and 0.7 have been frequently observed in the case of PdCu alloys [44-45; 77]. Gade et al. obtained an exponent of 0.583 for the permeation of pure hydrogen through a 16.7 µm thick $Pd_{59}Cu_{41}$ wt.% membrane produced by electroless plating [77]. Exponents higher than 0.5 in Sieverts's law indicate the increased influence of surface effects (adsorption, desorption, dissociation, recombination) in the hydrogen diffusion process through the membrane.

The permeance Q is temperature dependent according to an Arrhenius relationship:

$$Q = Q_0 e^{-E/RT}$$

Eq. (7.1-2)

with:

Q_0: pre-exponential factor ($mol.m^{-2}.s^{-1}.Pa^{-n}$);

E: activation energy for hydrogen permeation ($kJ.mol^{-1}$);

R: universal gas constant ($8.314\ J.mol^{-1}.K^{-1}$);

T: membrane temperature (K).

Using a ln-plot of the above-measured permeance value (slope of the Sieverts's law graph) versus 1/T, one can determine the activation energy required for hydrogen permeation through the bulk of the 50 µm $Pd_{60}Cu_{40}$ wt.% foil, as well as the pre-exponential factor Q_0 defined in Eq. (7.1-2). The result is presented in Fig. 7.1-2.

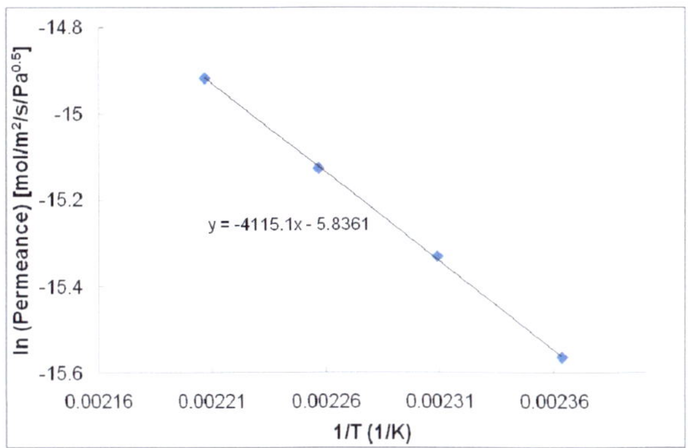

Fig. 7.1-2: Calculation of the permeance of the 50 µm $Pd_{60}Cu_{40}$ foil from the H_2 permeation measurements carried out above.

An activation energy E_a=34.2 kJ·mol^{-1}and a pre-exponential factor Q_0=2.92x10^{-3} mol.m^{-2}.s^{-1}.Pa$^{0.5}$ are calculated from Fig. 7.1-2 for this membrane.

The hydrogen permeability derived from the measured fluxes at 150°C and the nominal thickness of the foil is approx. 8.7x10^{-12} mol·m^{-1}·s^{-1}·Pa$^{-0.5}$. Pan et al. [44] published a detailed study of the hydrogen permeation through a 4 µm thick composite PdCu membrane with 66 wt.% Pd on a tubular ceramic support over a wide temperature range from room temperature up to 500°C. At 150°C their membrane showed a hydrogen permeability of 1.2x10^{-10} mol·m^{-1}·s^{-1}·Pa$^{-0.5}$, which is roughly 13 times higher than the permeability of the PdCu foil. The activation energy for hydrogen permeation differs as well: the PdCu foil showed a value of 34.2 kJ·mol^{-1}, whereas Pan et al. [44] reported a value of 21.3 kJ·mol^{-1} for their composite membrane. However, when analyzing these differences one has to be aware of the pronounced influence of the composition of the alloy on the hydrogen permeability especially in the PdCu system [44-45; 53-55]. In addition, the manufacturing process and the conditions employed, i.e., cold rolling versus electroless plating, also influences the permeation properties. Gade et al. reported a permeability of 1.33x10^{-8} mol.m^{-1}.s^{-1}.Pa$^{-0.5}$ at 400 °C for a 16.7 µm thick $Pd_{59}Cu_{41}$ wt.% film made by electroless plating [77]. Using the above mentioned activation energy and pre-exponential factor, the permeability of the 50 µm $Pd_{60}Cu_{40}$ foil at 400°C was calculated to be 3.24×10^{-10} $mol.m^{-1}.s^{-1}.Pa^{-0.5}$, which is 40 times less than that found by Gade.

7.1.2. PdCu composite membrane prepared by electroless plating

The hydrogen permeation results obtained with the 5 µm thick composite PdCu membrane (59 wt.% Pd) on a planar TRUMEM® substrate are provided in Fig. 7.1-3. The hydrogen fluxes measured at different temperatures are plotted vs. Sieverts's driving force with an exponent of 0.5. The permeation behavior shows a substantial scattering around Sieverts's law, with a substantial deviation towards higher exponents (closer to 1), compared to the well defined fluxes observed in the case of the 50 µm PdCu foil. The fact that the data points do not fit very well with Sieverts's equation could have different causes. The electroless deposition process does not guarantee that the exact alloy composition has been reached everywhere in the layer, even after the heat treatment. Inhomogeneities in the layer composition or Cu particles at the surface could influence the hydrogen diffusion. Dominant surface effects appearing in the case of low membrane thicknesses also bring permeation exponents close to 1; this, however, should not apply here.

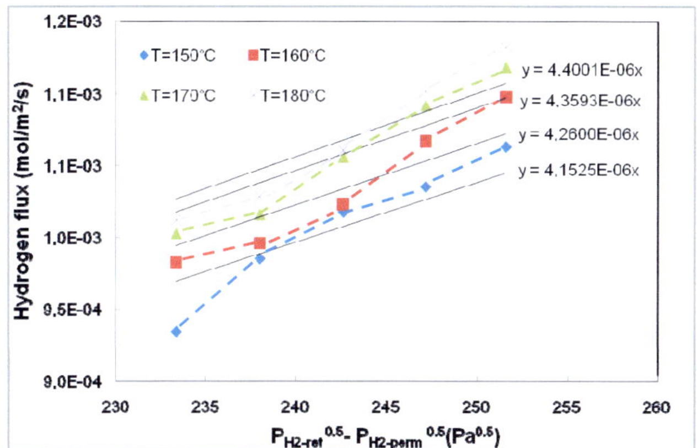

Fig. 7.1-3: H_2 permeation measurements carried out on a 5 µm thick $Pd_{60}Cu_{40}$ membrane produced by electroless plating. Influence of the temperature and partial pressure difference on the permeating flux.

The ln plot of the hydrogen permeance of the membrane vs. 1/T allows for the determination of the activation energy and pre-exponential factor in Fig 7.1-4.

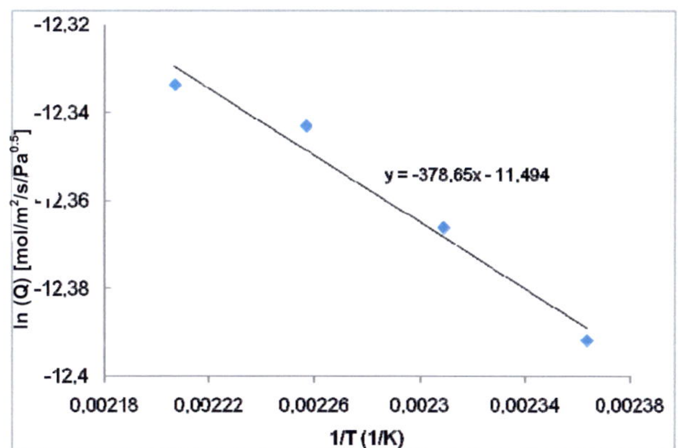

Fig. 7.1-4: Calculation of the permeance of the 5 µm $Pd_{59}Cu_{41}$ wt.% composite membrane produced by electroless plating on a TRUMEM® substrate.

The pre-exponential factor is 1.02×10^{-5} mol.m^{-2}.s^{-1}.Pa$^{-0.5}$ but with only 3.1 kJ/mol the apparent activation energy of the hydrogen permeation is unexpectedly low (34.2 kJ/mol were found in the case of the 50 µm Pd$_{60}$Cu$_{40}$ wt.% foil). This supports the hypothesis of some defects in the layer. The mass transport through defects is associated with a negative activation energy. The 3.1 kJ/mol measured could be the result of the combination of hydrogen transport though pores and transport through the film. The hydrogen fluxes observed here are, in any case, much lower than that through the large pores (0.2 µm) of the uncoated support, which eliminates a transport mechanism through pores only. Unfortunately, no nitrogen permeation measurements were carried out with this membrane, which would have made the calculation of the H$_2$/N$_2$ separation factor possible and thus a statement on the presence of defects. The use of the 50 µm PdCu foil was preferred for the hydroxylation experiments. Several aspects speak for their utilization in the reactor instead of the electroless plated composite membrane. The foils are a commercial product with a well defined composition and are much less susceptible to having inhomogeneities. This is illustrated by the excellent fit of the hydrogen transport through the foil to Sieverts's law. In addition to that, the less complex structure compared to the 3-layer composite membrane makes its surface more suitable (no pores and a more even state) for the deposition of the catalyst via magnetron sputtering.

7.1.3. PdCu sputtered with 1 µm PdAu

In the next sub-sections, the measured hydrogen permeance of the surface-modified membranes is presented. Please note that the activation energies and pre-exponential factors indicated correspond to those of both layers: the 50 µm thick PdCu support plus the sputtered top layer, which can be seen as a composite membrane. In a later section, the permeability of the individual sputtered layers is assessed and compared to that of the composite membrane. To isolate the coated layer, the composite membrane will be considered as a series of resistances to hydrogen permeation, where the flux values of the first resistance, the PdCu support, are known.

The hydrogen permeation properties of the surface modified foils have been measured using the same procedure as described earlier. A nitrogen sweep flow of 80 mL/min at the permeate side was used, the temperature was set at 150°C, 160°C and 170°C, respectively. The hydrogen partial pressure at the retentate side of the foil was gradually increased in the range of 1 to 3 bar. No homogenization heat treatment was carried out prior to the measurements due to the risk of the migration of Cu atoms into the PdAu top layer at high temperature, which would have modified the catalytic properties of the PdAu surface.

The $Pd_{60}Cu_{40}$ foil sputtered with 1 µm of $Pd_{90}Au_{10}$ shows the permeation behavior sketched in Fig. 7.1-5 measured at 3 different temperatures. A *n* exponent of 0.5 from Sieverts's law applies. Please note that a more limited number of experimental points have been measured here compared to other catalysts due to the late appearance of a hydrogen leakage through the seal while increasing the driving force.

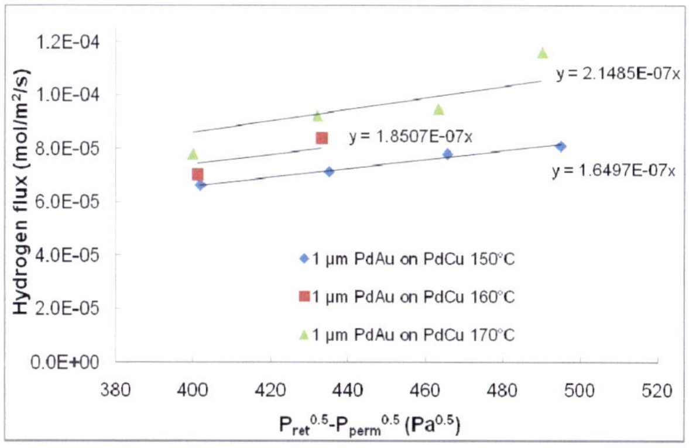

Fig. 7.1-5:   *H_2 permeation measurements carried out on a 50 µm thick $Pd_{60}Cu_{40}$ foil sputtered with 1 µm of $Pd_{90}Au_{10}$. Influence of the temperature and partial pressure difference on the permeating flux.*

Here again, Sieverts's law enables the calculation of the hydrogen permeance of the PdAu-sputtered PdCu foil. All sputtered PdCu foils are from the same production batch as the uncoated foil characterized in part 7.1.1 and are expected to obey the same permeation law. The 1 µm PdAu-coated PdCu foil logically shows lower hydrogen fluxes at the same temperature and driving force than the PdCu support on its own. The activation energy and pre-exponential factor for hydrogen permeation through the PdAu-PdCu composite membrane can be calculated from the data supplied in Fig. 7.1-5. The ln-plot of the permeance of the PdAu-sputtered PdCu foil vs 1/T can be seen in Fig 7.1-6.

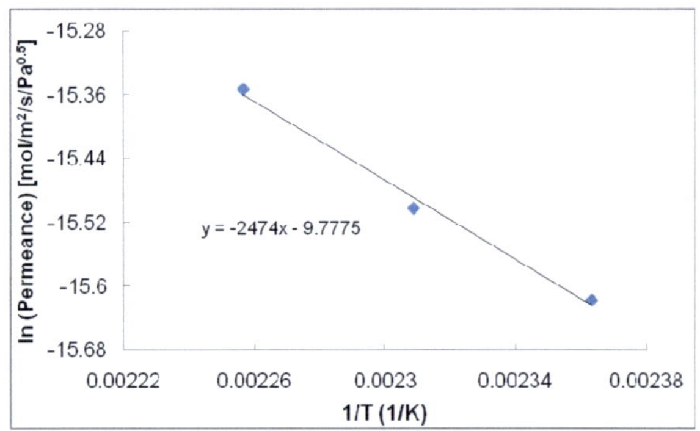

Fig. 7.1-6: Calculation of the permeance of the 1 µm Pd$_{90}$Au$_{10}$ sputtered Pd$_{60}$Cu$_{40}$ foil (as a bi-layer membrane) from the H$_2$ permeation measurements carried out above.

An activation energy of 20.5 kJ/mol and a pre-exponential factor Q_0=5.67x10^{-5} mol.m^{-2}.s^{-1}.Pa$^{-0.5}$ are obtained for the bi-layer PdAu-coated PdCu foil. Gade et al. have produced and characterized PdAu thin films made by electroless plating on a polished stainless steel support, from which they were subsequently separated for the purpose of measuring the hydrogen permeation of only the PdAu alloy [78]. An adequate comparison of the permeation results in this project with the PdAu membrane from Gade et al. will be done further on after the isolation of the coating via the resistance series model and the calculation of the permeability of the 1 µm PdAu sputtered layer on its own.

7.1.4. PdCu sputtered with V_2O_5/PdAu

Using the same experimental procedure for hydrogen permeation measurements, the V_2O_5/PdAu-sputtered foil was characterized at 4 different temperatures, ranging from 150°C to 180°C. Fig. 7.1-7 plots the permeation behavior according to Sieverts's law.

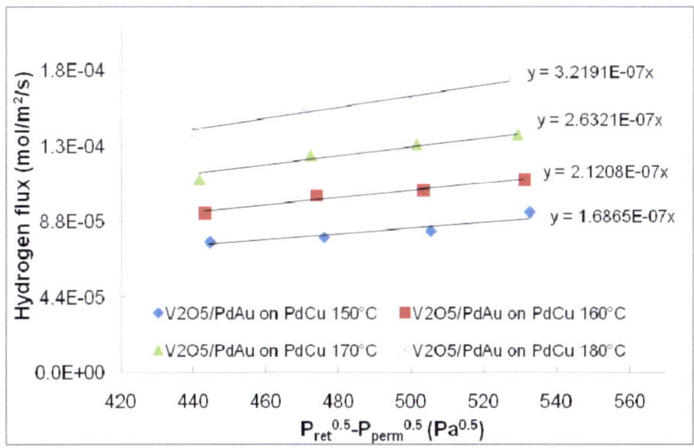

Fig. 7.1-7: H_2 permeation measurements carried out on a 50 µm thick $Pd_{60}Cu_{40}$ foil sputtered with V_2O_5 and $Pd_{90}Au_{10}$ particles. Influence of the temperature and partial pressure difference on the permeating flux.

As a first observation, one notices very similar fluxes to the unmodified PdCu foil, which is an indication that the briefly sputtered catalyst has not extensively modified the hydrogen transport mechanism through the PdCu support. The objective of this catalytic modification was to reach an interrupted insular pattern of V_2O_5 particles supporting PdAu particles, applied in a second step, in order to increase the surface area of the PdAu top catalyst.

The parameters of the Arrhenius law governing the hydrogen permeance of the composite membrane can be gleaned from Fig 7.1-8 below.

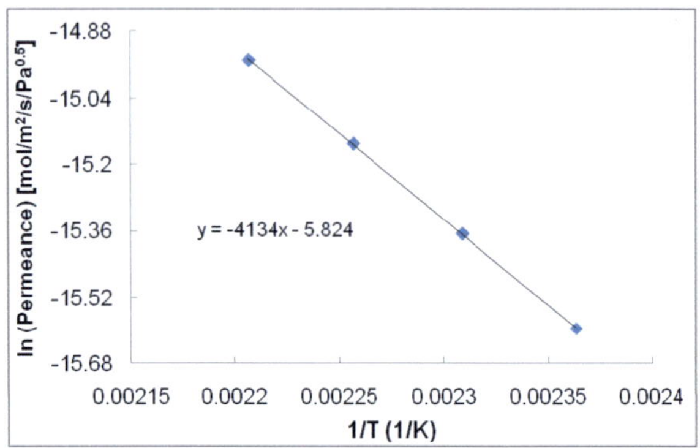

Fig. 7.1-8: Calculation of the permeance of the V_2O_5/ $Pd_{90}Au_{10}$ sputtered $Pd_{60}Cu_{40}$ foil (as a bi-layer membrane) from the H_2 permeation measurements carried out above.

A pre-exponential factor Q_0=2.96x10^{-3} mol.m^{-2}.s^{-1}.Pa$^{-0.5}$ and an activation energy of 34.4 kJ/mol are measured. This activation energy is quasi identical to the one of the PdCu support (34.2 kJ/mol) as are the pre-exponential factors (2.92x10^{-3} mol.m^{-2}.s^{-1}.Pa$^{-0.5}$ was obtained with PdCu). This confirms the low influence of the thin catalyst produced on the support during the short sputtering period. The insular structure of the vanadium oxide catalyst was observed by EDX analysis. The uniform aspect of the catalyst surface tends to indicate, despite the higher roughness, that the PdAu sputtered catalyst has formed a thin layer instead of particles on the V_2O_5 phase. A flux comparison between all surface-modified membranes under the same process conditions is supplied in a later section.

7.1.5. PdCu sputtered with 5 µm PdGa

In the case of the PdGa catalyst, 3 distinct PdCu foils have been sputtered with different layers of PdGa, of nominal thicknesses 5 µm, 1.4 µm and 0.4 µm respectively. The reason behind the preparation of several thicknesses of this catalytic layer is the much lower hydrogen permeation rate noticed using the 5 µm PdGa-coated foil (which was initially prepared), which strongly limits the hydrogen partial pressure that can be reached in the reactor during the hydroxylation experiments. The performance of the PdGa catalyst could thus not be fully assessed and compared to the other systems due to the low hydrogen supply through the membrane. Thinner PdGa layers had to be prepared.

The 5 µm thick PdGa sputtered foil was initially characterized using the same experimental procedure. Its permeation behavior is represented in Fig. 7.1-9. No diffusion heat treatment was used as for palladium-gold (original catalyst composition maintained and with it the expectations for the benzene hydroxylation reaction).

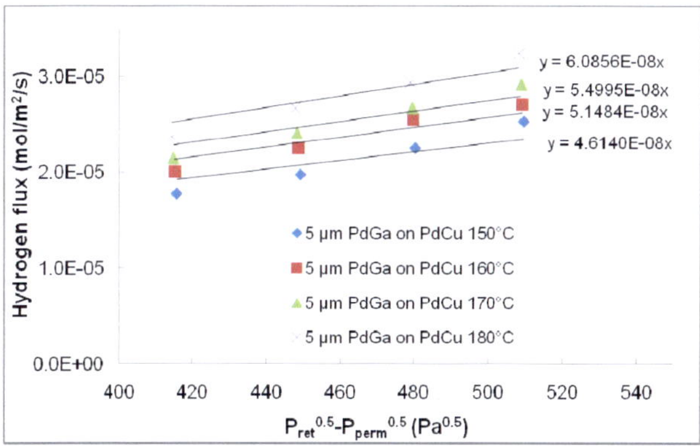

Fig. 7.1-9: H_2 permeation measurements carried out on a 50 µm thick $Pd_{60}Cu_{40}$ foil sputtered with 5 µm of $Pd_{50}Ga_{50}$. Influence of the temperature and partial pressure difference on the permeating flux.

With 5 µm PdGa sputtered on a 50 µm PdCu foil, much lower hydrogen fluxes are measured through the bi-layer membrane in comparison to those through the PdCu foil under the same process conditions. A H_2-flux of about 2.3×10^{-5} mol/m²/s is measured through the composite membrane at 150°C and 500 $Pa^{0,5}$ of driving force, against 8.7×10^{-5} mol/m²/s through PdCu. No comparative data could be found in the literature for this type of layer. The characterization of the catalytic layer showed the presence of different phases in the alloy (the PdGa intermetallic phase but also probably the Pd_2Ga phase – cf previous XRD results), which are not known for their hydrogen permeation properties. The permeation behaviour through such a layer is probably dominated by transport through the pure α-Pd phase. However, this assumption cannot be verified as the different phases cannot be isolated from one another in the layer and as a 100% intermetallic $Pd_{50}Ga_{50}$ layer could not be produced by sputtering from the target of this composition.

Fig. 7.1-10 enables the calculation of the activation energy and pre-exponential factor of the 5 µm PdGa-PdCu membrane.

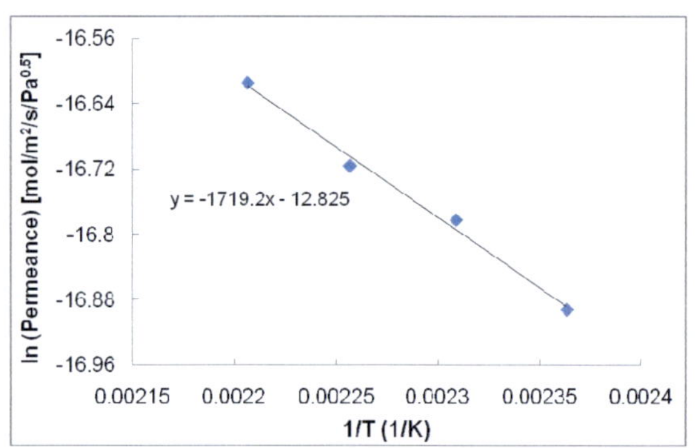

Fig. 7.1-10:   Calculation of the permeance of the 5 µm $Pd_{50}Ga_{50}$ sputtered $Pd_{60}Cu_{40}$ foil (as a bi-layer membrane) from the H_2 permeation measurements carried out above.

An activation energy E_a=14.3 kJ/mol and a pre-exponential factor Q_0=2.69x10^{-6} mol.m^{-2}.s^{-1}.Pa$^{-0.5}$ were measured. This activation energy is more than two times lower than that of the unmodified foil and 2/3 of the value for the PdAu/PdCu foil. A lower activation energy could be caused by the transport at the interface, limiting the diffusion through the membrane. It might also be due to the intermetallic PdGa phase, which does not allow for a proper hydrogen transport. The desired hydrogen transport may be possible in a Pd-richer phase in the sputtered layer.

7.1.6. PdCu sputtered with 1.4 µm PdGa

The foil sputtered with a thinner PdGa layer of 1.4 µm was characterized by applying the usual procedure. The hydrogen permeation obtained via this is illustrated in Fig 7.1-11. Intermediate fluxes between those obtained through the PdAu-sputtered foil and the 5 µm PdGa-sputtered PdCu foil were observed, with permeation values ranging from 1.01×10^{-7} mol.m^{-2}.s^{-1}.Pa$^{-0.5}$ at 150°C to 1.41×10^{-7} mol.m^{-2}.s^{-1}.Pa$^{-0.5}$ at 180°C. Compared to the slightly thinner PdAu coating (permeance of 1.65×10^{-7} mol.m^{-2}.s^{-1}.Pa$^{-0.5}$ at 150°C), the 1.4 µm PdGa layer appears less permeable. The activation energy value confirms this trend. The ln-plot of the permeance vs. 1/T is shown in Fig. 7.1-12.

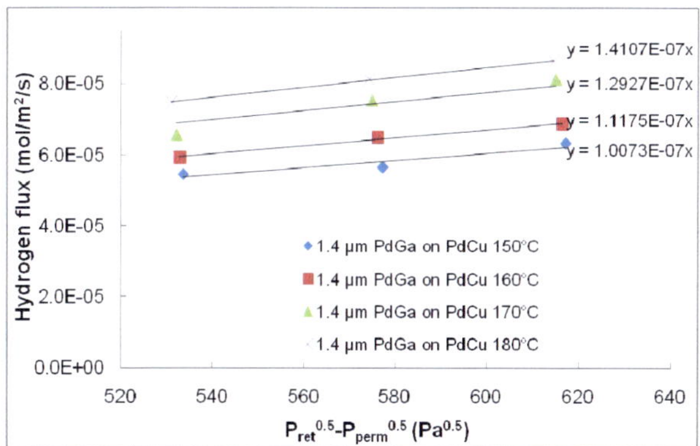

Fig. 7.1-11: H₂ permeation measurements carried out on a 1.4 µm PdGa sputtered 50 µm Pd$_{60}$Cu$_{40}$ foil. Influence of the temperature and partial pressure difference on the permeating flux.

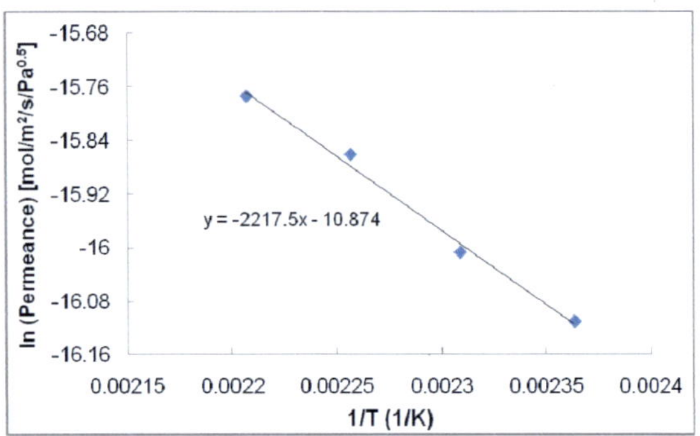

Fig. 7.1-12: Calculation of the permeance of the 1.4 µm Pd$_{50}$Ga$_{50}$ sputtered Pd$_{60}$Cu$_{40}$ foil (as a bi-layer membrane) from the H$_2$ permeation measurements carried out above.

The activation energy in this case is 18.4 kJ/mol and the pre-exponential factor is 1.89×10^{-5} mol.m^{-2}.s^{-1}.Pa$^{-0.5}$. It is 4.1 kJ/mol greater than that of the 5 µm PdGa/PdCu membrane, which is less permeable to H$_2$ due the higher PdGa thickness. The 1 µm PdAu-modified PdCu foil, whose activation energy is 2.1 kJ/mol higher, is approx. 50% more hydrogen permeable than the 1.4 µm PdGa/PdCu membrane. The permeance values and activation energy of the composite membranes show a variation in opposite ways. This is also observed with the thinner PdGa layer.

7.1.7. PdCu sputtered with 0.4 µm PdGa

Fig. 7.1-13 illustrates the permeated fluxes measured at the different driving forces and temperatures for the 0.4 µm PdGa sputtered PdCu foil. The hydrogen fluxes obtained are, similarly to those through the briefly sputtered V$_2$O$_5$/PdAu system on PdCu, very close to the fluxes of the uncoated PdCu support (range of 6.0×10^{-5} mol/m^2/s -1.6×10^{-4} mol/m^2/s for the applied parameters). It appears, in this case, that the influence of the catalyst on the hydrogen permeation through the composite membrane is minimal.

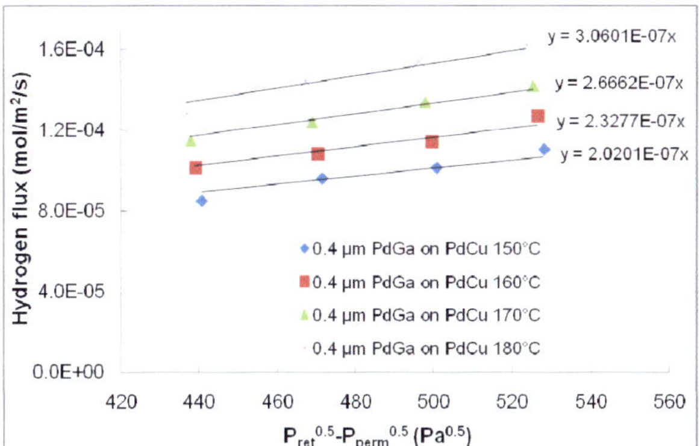

Fig. 7.1-13: H_2 permeation measurements carried out on a 0.4 µm PdGa sputtered 50 µm $Pd_{60}Cu_{40}$ foil. Influence of the temperature and partial pressure difference on the permeating flux.

Fig. 7.1-14 shows the permeance parameters.

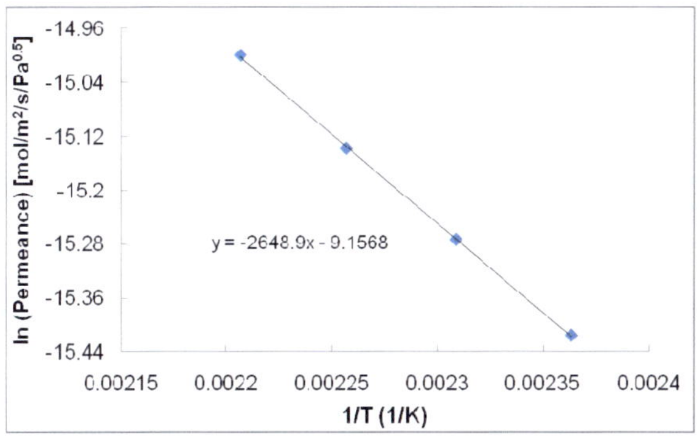

Fig. 7.1-14: Calculation of the permeance of the 0.4 µm $Pd_{50}Ga_{50}$ sputtered $Pd_{60}Cu_{40}$ foil (as a bi-layer membrane) from the H_2 permeation measurements carried out above.

An activation energy E_a=22.0 kJ/mol and Q_0=1.05x10^{-4} mol.m^{-2}.s^{-1}.Pa$^{-0.5}$ are measured for this bi-layer membrane. The activation energy seem to increase proportionally to the decrease in the thickness of the deposited PdGa layer.

14.3 kJ/mol, 18.4 kJ/mol and 22.0 kJ/mol were measured through 5 µm, 1.4 µm and 0.4 µm of PdGa deposited on PdCu respectively.

Table 7.1-1 summarizes the values of the activation energies and pre-exponential factors measured after PdGa, PdAu and V_2O_5/PdAu-modification of the PdCu foils respectively. The data obtained for the unmodified $Pd_{60}Cu_{40}$ foil is also given.

Table 7.1-1: Values of activation energy and permeance pre-exponential factors measured for hydrogen permeation through the different catalytic systems

Type of foil modification	Activation energy E_a measured for H_2 permeation (kJ/mol)	Pre-exponential factor Q_0 (mol/m^2/s/Pa$^{0.5}$)
unmodified $Pd_{60}Cu_{40}$ foil	34.2 ± 0.3	2.92×10^{-3}
5 µm $Pd_{50}Ga_{50}$ on $Pd_{60}Cu_{40}$	14.1 ± 0.1	2.16×10^{-7}
1.4 µm $Pd_{50}Ga_{50}$ on $Pd_{60}Cu_{40}$	18.4 ± 0.2	1.89×10^{-5}
0.4 µm $Pd_{50}Ga_{50}$ on $Pd_{60}Cu_{40}$	22.0 ± 0.2	1.05×10^{-4}
1 µm $Pd_{90}Au_{10}$ on $Pd_{60}Cu_{40}$	20.5 ± 0.2	5.67×10^{-5}
V_2O_5/$Pd_{90}Au_{10}$ on $Pd_{60}Cu_{40}$	34.4 ± 0.3	2.96×10^{-3}

7.1.8. Flux comparison between surface-modified membranes

The fluxes measured for different composite membranes are compared with each other at the same temperature. Due to their flux values being very close to those through the unmodified PdCu support, the results for 0.4 µm PdGa and V_2O_5/PdAu have been excluded from Fig. 7.1-15, Fig 7.1-16 and Fig. 7.1-17, which show the permeances at 150°C, 160°C and 170°C respectively. No relevant calculation on the effect of the deposited layer can be realized if the sputtered foils show the same permeation as the support at all temperatures. The flux comparison was realized between the PdCu support (dashed line in Fig. 7.1-15, Fig 7.1-16 and Fig. 7.1-17) and the PdAu, 1.4 µm PdGa and 5 µm PdGa-sputtered PdCu foils. The V_2O_5/PdAu and 0.4 µm PdGa systems are too thin to influence the hydrogen permeance through the whole membrane and thus do not provide information on the hydrogen transport through them; they will, however, affect the hydroxylation reaction on benzene as it takes place on the catalyst surface.

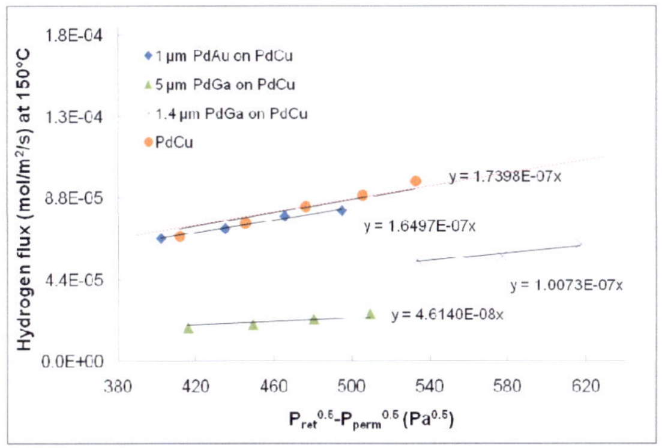

Fig. 7.1-15: *Comparison of the H_2 fluxes through different surface-modified PdCu foils at T=150°C.*

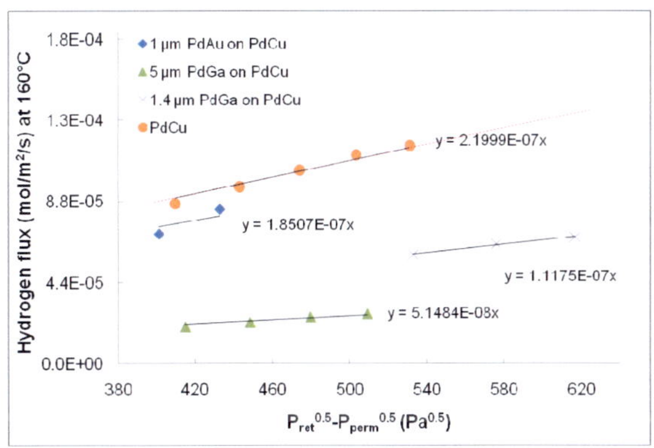

Fig. 7.1-16: *Comparison of the H_2 fluxes through different surface-modified PdCu foils at T=160°C.*

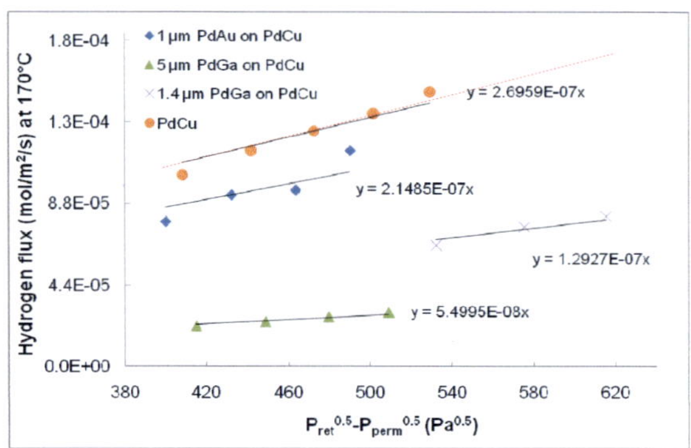

Fig. 7.1-17: Comparison of the H_2 fluxes through different surface-modified PdCu foils at T=170°C.

By comparing the permeance of the different membranes at the same temperature, one can observe the resistance to hydrogen transport produced by the coated catalyst with respect to the PdCu support on its own. Several observations can also be made in relation to the effect of the temperature on the hydrogen permeation through the modified systems.

In Fig. 7.1-15 to Fig. 7.1-17, one observes that the temperature plays a higher role in increasing the flux through the PdCu foil than through the other systems. This directly correlates with the higher activation energy calculated for the PdCu support. The hydrogen transport at 170°C is less promoted through the surface-modified systems, as is shown in Table 7.1-2, which indicates the % of initial permeance reached at the different temperatures. The reference is the PdCu support without modification.

Table 7.1-2: Hydrogen permeance measured at T=170°C for the different systems and comparison with that through the unmodified $Pd_{60}Cu_{40}$ foil

Type of catalytic modification	H_2 permeance (mol/m^2/s/Pa$^{0.5}$) at T=150°C:	% of the H_2 permeance of the PdCu foil
unmodified $Pd_{60}Cu_{40}$ foil	1.74×10^{-7}	100
5 µm $Pd_{50}Ga_{50}$ layer on $Pd_{60}Cu_{40}$	4.61×10^{-8}	26.5
1.4 µm $Pd_{50}Ga_{50}$ layer on $Pd_{60}Cu_{40}$	1.01×10^{-7}	58.0
1 µm $Pd_{90}Au_{10}$ layer on $Pd_{60}Cu_{40}$	1.65×10^{-7}	94.8
	at T=160°C:	
unmodified $Pd_{60}Cu_{40}$ foil	2.20×10^{-7}	100
5 µm $Pd_{50}Ga_{50}$ layer on $Pd_{60}Cu_{40}$	5.15×10^{-8}	23.4
1.4 µm $Pd_{50}Ga_{50}$ layer on $Pd_{60}Cu_{40}$	1.12×10^{-7}	50.9
1 µm $Pd_{90}Au_{10}$ layer on $Pd_{60}Cu_{40}$	1.85×10^{-7}	84.1
	at T=170°C	
unmodified $Pd_{60}Cu_{40}$ foil	2.70×10^{-7}	100
5 µm $Pd_{50}Ga_{50}$ layer on $Pd_{60}Cu_{40}$	5.50×10^{-8}	20.4
1.4 µm $Pd_{50}Ga_{50}$ layer on $Pd_{60}Cu_{40}$	1.29×10^{-7}	47.8
1 µm $Pd_{90}Au_{10}$ layer on $Pd_{60}Cu_{40}$	2.15×10^{-7}	79.6

At 170°C, the 1 µm-PdAu sputtered PdCu foil shows a permeance, which is 79.6% that of the support (2.70×10^{-7} mol.m^{-2}.s^{-1}.Pa$^{-0.5}$), whereas it reached 94.8% of its permeance at a temperature, which is 20°C lower.

The same is observed with the other PdGa systems, which show a higher transport resistance to hydrogen than PdAu, particularly as the layer thickness increases. The 1.4 µm PdGa sputtered membrane delivers a flow, which is 58% of that through its support at 150°C for the same driving force, whereas a 5 µm PdGa deposited film only results in 26.5% of the initial flow. These values are further reduced as the temperature increases, down to 20.4% for 5 µm PdGa at 170°C.

The flux through the 1 µm PdAu-sputtered foil as it appears in Fig. 7.1-15 has to be excluded from the permeability calculation of the isolated layer due to the hydrogen flux at 150°C, which is in the range of that through the uncoated support.

7.1-9: Isolation of the permeability of the sputtered layers using a resistance series model

If one considers the composite membrane as a superposition of 2 membranes of different thicknesses and compositions, which are perfectly adhered to each other (with no other resistance to hydrogen transport at the interface than the lattice transition; total absence of defects or pinholes), then it is possible to decompose it as a series of 2 membranes showing the same hydrogen flux (conservation of the flux between both sides of the isolated membranes), as is illustrated in Fig. 7.1-18.

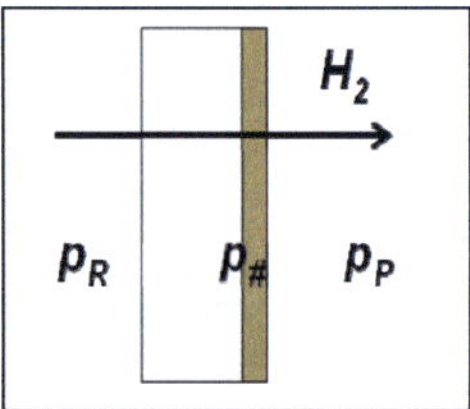

Fig. 7.1-18: *Scheme of hydrogen permeating through a bi-layer membrane as used in the resistance series model*

In this case, it is possible to isolate the permeability of the sputtered layer on its own due to the expression of the flux through the different layers (substrate and coating).

1) $\quad J_{PdCu} = Q_{PdCu}\left(p_R^{n_{PdCu}} - p_\#^{n_{PdCu}}\right)$ Eq (7.1-4)

2) $\quad J = Q\left(p_\#^{n} - p_P^{n}\right)$ Eq (7.1-5)

3) $\quad J_{PdCu} = J$ Eq (7.1-6)

$$\textit{from } 1): \quad p_\# = \left(p_R^{n_{PdCu}} - \frac{J}{Q_{PdCu}}\right)^{\left(\frac{1}{n_{PdCu}}\right)}$$

Eq (7.1-7)

$$\textit{in } 2): \quad J = Q\left(\left[p_R^{n_{PdCu}} - \frac{J}{Q_{PdCu}}\right]^{\left(\frac{1}{n_{PdCu}}\cdot n\right)} - p_P^{n}\right)$$

Eq (7.1-8)

$$\Rightarrow Q = \frac{J}{\left(p_R^{n_{PdCu}} - \dfrac{J}{Q_{PdCu}}\right)^{\left(\frac{n}{n_{PdCu}}\right)} - p_P^{n}}$$

Eq (7.1-9)

$$\textit{for } n = n_{PdCu} : Q = \frac{J}{p_R^{n} - p_P^{n} - \dfrac{J}{Q_{PdCu}}}$$

Eq (7.1-10)

The exponent which applies for the driving force of the composite membranes and the PdCu support was identical for all catalytic systems with a value of 0.5 according to Sieverts's law. One assumes that the same n exponent applies to the support and the top deposited layer. This leads to the expression of the permeance Q of the deposited catalyst on its own provided by Eq. (7.1-10). Using Eq. (7.1-10), one can calculate the isolated permeance of the top catalytic layer for the systems, which have layers in the appropriate thickness (1 µm PdAu, 1.4 µm and 5.0 µm PdGa), so that the flux differs from that of the PdCu support. The results are presented in Table 7.1-4 and compared to PdCu.

Table 7.1-3: Permeances of the PdAu and PdGa sputtered foils and of the catalytic layers on their own obtained with the resistance series model

Membrane type	T(°C)	Permeance of the composite membrane (mol/m^2/s/Pa$^{0.5}$)	Re-calculated permeance of the layer on its own (mol/m^2/s/Pa$^{0.5}$)
PdCu	150	1.74x10^{-7}	/
	160	2.20x10^{-7}	/
	170	2.69x10^{-7}	/
	180	3.32x10^{-7}	/
1 µm PdAu on PdCu	160	1.85x10^{-7}	1.19x10^{-6}
	170	2.15x10^{-7}	1.08x10^{-6}
5 µm PdGa on PdCu	150	4.61x10^{-8}	6.23x10^{-8}
	160	5.15x10^{-8}	6.68x10^{-8}
	170	5.50x10^{-8}	6.87x10^{-8}
	180	6.08x10^{-8}	7.40x10^{-8}
1.4 µm PdGa on PdCu	150	1.01x10^{-7}	2.42x10^{-7}
	160	1.12x10^{-7}	2.28x10^{-7}
	170	1.29x10^{-7}	2.49x10^{-7}
	180	1.41x10^{-7}	2.46x10^{-7}

As 2 PdGa coatings have been isolated, a relevant parameter to be calculated is the permeability of the layers, which is the permeance multiplied by the layer thickness, as indicated by Eq. (7.1-11).

$$Q' = h_{layer} \times Q \qquad \text{Eq. (7.1-11)}$$

with Q' the permeability expressed in mol/m/s/Pa^n, h_{layer} the thickness of the H$_2$ permeable layer and Q its permeance in mol/m^2/s/Pa^n.

As they are made of the same alloy, the 2 PdGa films are expected to feature the same permeability value.

The calculation of the permeability from the permeance data in Table 7.1-3: and the plot of ln(permeability) vs. 1/T (Fig. 7.1-19) enables the determination of the activation energies and pre-exponential factors of the isolated layers as shown in Table 7.1-4.

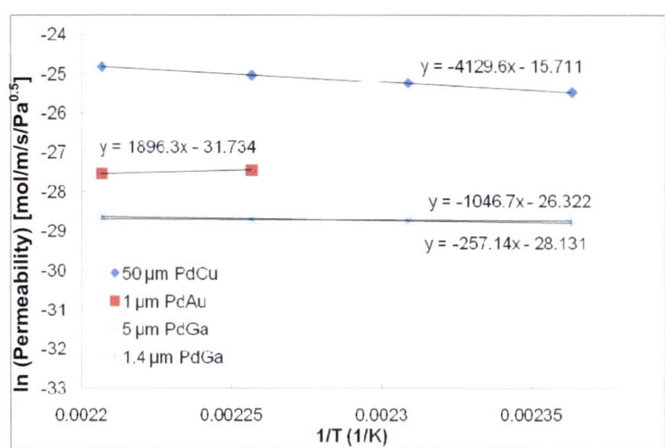

Fig. 7.1-19: ln(permeability) vs. 1/T plot for determination of the permeability constants of the different layers.

One observes the highest permeability with the PdCu foil, which is about an order of magnitude higher than the PdAu layer and 2 orders higher than the PdGa layers. As expected, both the 1.4 µm PdGa and 5 µm PdGa layers have a quasi-identical permeability ranging from 3.2×10^{-13} mol.m^{-1}.s^{-1}.Pa$^{-0.5}$ to 3.7×10^{-13} mol.m^{-1}.s^{-1}.Pa$^{-0.5}$ between 150°C and 180°C.

Table 7.1-4: Values of activation energy and permeability pre-exponential factors measured for hydrogen permeation through bi-layer composite membranes and isolated sputtered layers

Membrane type	Activation energy E_a measured for H_2 permeation (kJ/mol)	Pre-exponential factor Q'_0 (mol/m/s/Pa$^{0.5}$)
50 µm Pd$_{60}$Cu$_{40}$ foil	34.2	1.50×10^{-7}
5 µm Pd$_{50}$Ga$_{50}$ on Pd$_{60}$Cu$_{40}$	14.3	1.35×10^{-11}
5 µm Pd$_{50}$Ga$_{50}$ alone	5.4	1.50×10^{-12}
1.4 µm Pd$_{50}$Ga$_{50}$ on Pd$_{60}$Cu$_{40}$	18.4	2.65×10^{-11}
1.4 µm Pd$_{50}$Ga$_{50}$ alone	5.4	1.50×10^{-12}
1 µm Pd$_{90}$Au$_{10}$ on Pd$_{60}$Cu$_{40}$	20.5	5.67×10^{-11}
1 µm Pd$_{90}$Au$_{10}$ alone	-15.7	1.65×10^{-14}

The average activation energy measured for H_2 permeation through the PdGa layers PdGa is 5.4 kJ/mol and the pre-exponential factor is 1.5×10^{-12} mol.m^{-1}.s^{-}

^{1}.Pa$^{-0.5}$. Surprisingly, the activation energy for the isolated PdAu layer is negative, with a value of -15.7 kJ/mol. The SEM and XRD analysis of the PdAu-coated membrane have not revealed the presence of defects or discontinuity in the layer, which would lead to a transport through pores and which is associated with a negative activation energy.

This negative value tends to indicate that the resistance series model used for the bi-layer membrane (with virtual intermediary partial pressure) is probably too simple for a satisfactory representation of the real permeation behavior, as the situation at the interface cannot be properly described. A more detailed model, including a contact resistance at the interface, would provide higher permeability values for the top layers, which would all show, as expected, positive activation energy values.

Gade et al. prepared and tested several self-supported PdAu alloy membranes, made by electroless plating, with different gold content and thicknesses [78]. One of them, a 9 wt.% Au with a 10.3 µm thick membrane showed a hydrogen permeance of 6.64x10^{-4} mol.m^{-2}.s^{-1}.Pa$^{-0.5}$ at 400°C with a high permselectivity H_2/N_2 of 1670. The presence of defects influencing the transport mechanism can, therefore, be excluded here. At 150°C, a permeance of 5.04x10^{-6} mol.m^{-2}.s^{-1}.Pa$^{-0.5}$ is obtained by isolating the 1.0 µm $Pd_{90}Au_{10}$ wt.% layer; this is 2 orders of magnitude below the value reported by Gade et al, however for a much lower temperature. If the layer permeabilities were compared, the 1 µm $Pd_{90}Au_{10}$ wt.% film would be only 1 order of magnitude less permeable at 250°C. Opposite behaviors regarding the temperature influence on the permeance were also found. This supports the hypothesis of an inadequate model, which results in a negative activation energy for the 1 µm $Pd_{90}Au_{10}$ layer.

Due to the specificity of the catalytic layers employed in this project, it has not been possible to find comparative permeation data in published articles to correlate the obtained measurements and observations on PdGa.

7.2. Membranes for oxygen dosage:

7.2.1. Ag foil

The results of oxygen permeation measurements with a 15 µm thick silver foil are presented in Fig. 7.2-1 Synthetic air served as the oxygen source on the retentate side. The temperature range of 150°C to 180°C was covered.

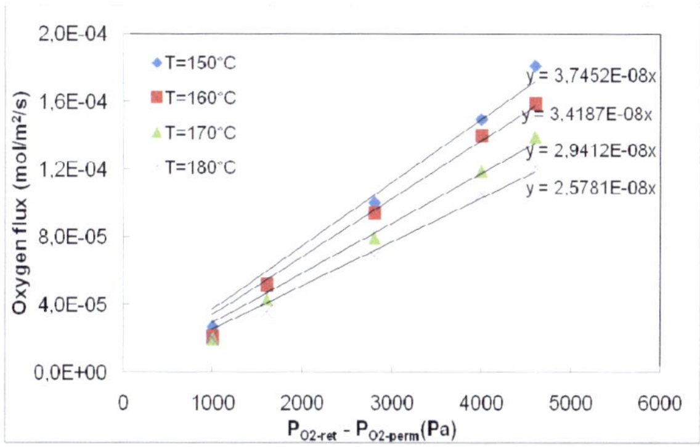

Fig.7.2-1: O_2 permeation measurements carried out on a 15 µm thick Ag foil. Influence of the temperature and the pressure difference on the permeation flux.

The experimental data reveal a linear relation between the permeated oxygen flow and the oxygen partial pressure difference:

$$J_{O_2} = Q \cdot \left(P_{O_2,ret} - P_{O_2,perm} \right) \qquad \text{Eq. (7.2-1)}$$

This behavior is an indication of flow of molecular oxygen through defects in the foil rather than transport of oxygen atoms through the metal lattice, which would be described by a Sieverts's type law where the driving force is the partial pressure difference at the exponent 0.5. Further evidence for the dominance of such rather small defects can be found in the temperature dependence of the oxygen flux; whether the transport is governed by a Knudsen diffusion mechanism or by a viscous flow through large pores can be determined from this. Knudsen diffusion occurs when the gas mean free path of the molecule is large when compared to the characteristic pore dimension; this means that the collision of the gas molecules with the pore walls dominates as opposed to the collision of the molecules with each over in the transport mechanism (the mean free path of the molecules can be estimated using λ = 100 nm/p, where p is the pressure in bar). The permeance is shown to be proportional to T^{-n}, a slightly better correlation coefficient is obtained for n=0.5 (temperature dependence in the case of Knudsen diffusion) in comparison to n=1.5 (viscous flow through large pores) [81]. Both trends are plotted in Fig. 7.2-2, where Knudsen diffusion through small pores appears as the dominating mechanism.

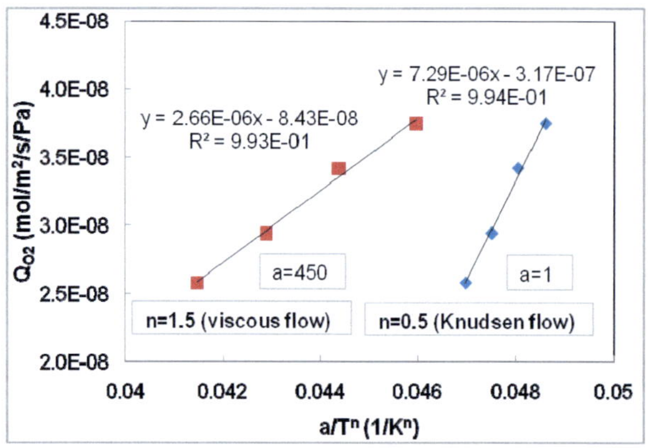

Fig. 7.2-2: Plot of the oxygen permeance of the 15 µm Ag foil produced vs the temperature $1/T^n$, with n=0.5 (Knudsen flow dependence) and n=1.5 (viscous flow).

The temperature dependence of the permeance points to the role of Knudsen diffusion and viscous flow in the permeation mechanism governing the oxygen transport through the foil. In this case, the prerequisite for Knudsen diffusion is a small pore size of 100 nm or less.

Oxygen flow through defects are not detrimental for the concept of distributed dosing of oxygen along the reactor as long as a relatively large number of small defects exists which are equally distributed over the whole membrane. The surface of the Ag foil was inspected by SEM to answer this question. The analysis showed the presence of many rolling traces and randomly distributed small surface defects, as illustrated in Fig. 7.2-3.

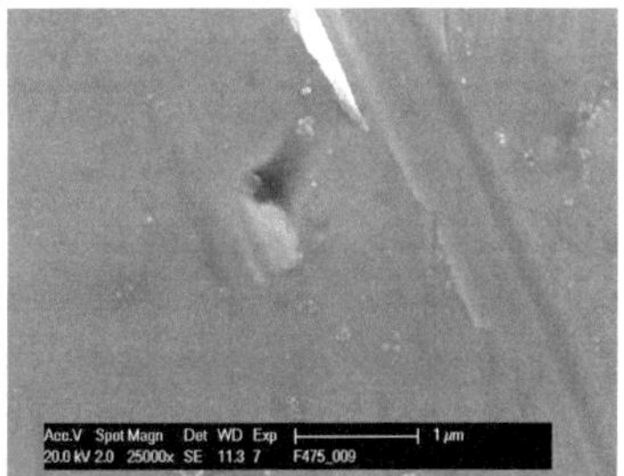
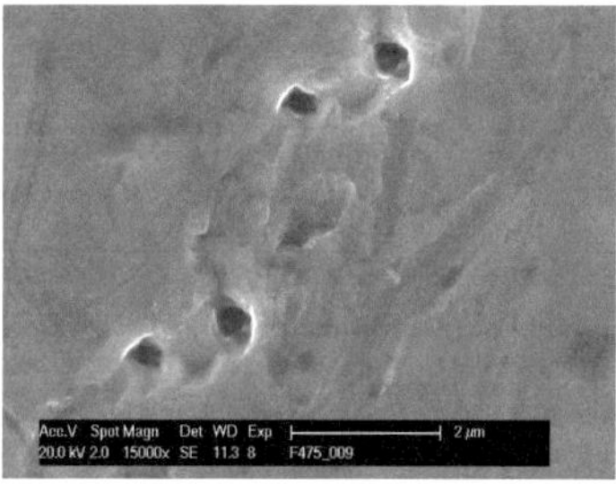

Fig.7.2-3: SEM images of defects on the surface of the 15 µm thick Ag foil.

These surface defects suggest a pore size around 100 nm in the bulk of the foil (expected pore size, which will enable Knudsen transport); their size, however, cannot be determined with accuracy. In addition to Fig. 7.2-2, the observed foil surface tends to indicate that the oxygen transport mechanism is influenced by both Knudsen transport through small defects and a viscous component of the flow through larger defects. No clear trend can be gleaned from this analysis.

7.2.2 Ag composite membrane

As for the Ag foil, the oxygen transport properties of the 5 µm Ag composite membrane produced on the TRUMEM® support were tested using synthetic air. The air pressure at the retentate side was increased to set the driving force for temperatures ranging from 150°C to 180°C. The absolute pressure in the central channel was 1 bar.

Fig. 7.2-4 shows the results of the permeation measurements. Note that the flux values are plotted with respect to oxygen partial pressure difference at the exponent 1.

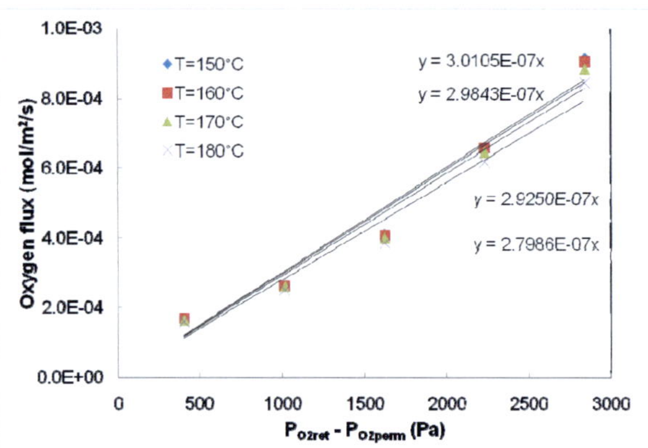

Fig.7.2-4: O_2 permeation measurements carried out on a 5 µm thick Ag membrane. Influence of the temperature and the pressure difference on the permeation flux.

Large oxygen fluxes in the range of 10^{-4} -10^{-3} mol/m²/s are obtained in the range of pressures covered. These fluxes exceed by 4 to 5 orders of magnitude the flux prediction made from Gryaznov data [46], which shows that the pores of the support might have not been sealed during the layer deposition by electroless plating. The evolution of the oxygen flux with the temperature shows a light decrease at higher temperatures. This suggests that defects participate to the oxygen transport, in non-dissociated form. Knudsen diffusion through small pores leads to a proportionality of the permeance with $1/T^{0.5}$, while a viscous flow through large pores leads to a permeance dependence with $1/T^{1.5}$. The analysis of the temperature dependence once again provides information on the dominating flow mechanism.

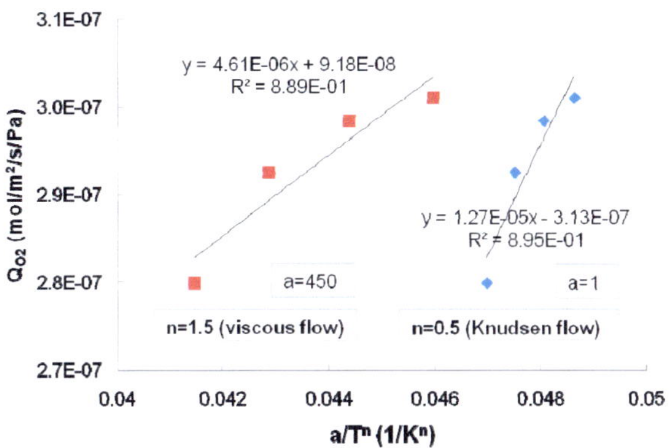

Fig. 7.2-5: Plot of the oxygen permeance of the 5 µm Ag membrane produced by electroless plating vs the the temperature $1/T^n$, with n=0.5 (Knudsen flow dependence) and n=1.5 (viscous flow).

Given the high scattering of the experimental points around the trend line shown in Fig. 7.2-5, in comparison to Fig. 7.2-2, and the small difference in the correlation coefficient, it was concluded that both the viscous flow and Knudsen diffusion influence the oxygen transport through the membrane. Due to the fact that permeating nitrogen was also measured here (permeate sweep flow of N_2 excluded here), it can be assumed that the pores may be somewhat larger in the 5 µm electroless plated Ag membrane than in the 15 µm Ag foil.

Due to imperfections in the structure of the 5 µm plated silver layer and to the unexpected permeation behavior observed, the membrane produced does not have the selective properties expected for the oxygen dosage. No compact membrane could be produced here. It features defects of unknown geometries and locations which renders the oxygen dosage imprecise. Its utilization for oxygen dosage in the reactor is, therefore, not recommended.

All the 0.1 µm pores of the TiO_2 top layer of the TRUMEM® support probably could not be covered by a 5 µm coating deposited by electroless plating. A (much) thicker coating is required to obtain a defect-free membrane. Further depositions to reach this state were not performed as the alternative of using the dense foil already existed.

7.2.3. UF-Trumem membrane

The UF-membrane produced by the sol-gel method on the TRUMEM® support was also tested in a temperature range between 150°C and 170°C. The air overpressure was moderately increased at the retentate side of the membrane. This membrane is not based on a dense layer and has a higher response to pressure changes.

Fig. 7.2-6 illustrates the oxygen fluxes obtained at the difference temperatures and the oxygen partial pressure difference.

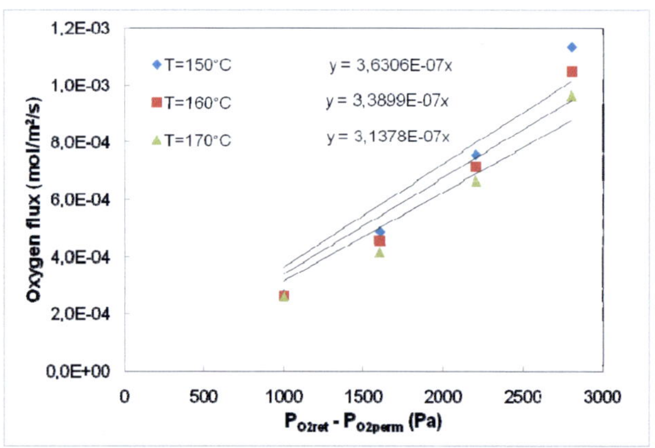

Fig.7.2-6: O_2 permeation measurements carried out on a UF-TRUMEM membrane. Influence of the temperature and the pressure difference on the permeation flux.

As observed in Fig. 7.2-6, the oxygen flux dependence on the pressure difference appears, in this case, to be more quadratic than linear. As expected, the fluxes measured here were quite high, in the range 10^{-4} mol/m²/s, for small pressure increases; this is due to the porosity of the top layer (30 nm is the expected size of the pores after the sol-gel process). The permeances obtained are an order of magnitude higher than those obtained through the 15 µm Ag foil, which featured defects. The permeation behavior once again featured decreasing fluxes for temperature increases. This is again seen as an indication for a flow controlled by Knudsen diffusion, which applies for gas transport through pores in above-mentioned diameter range.

The nitrogen flow measured by the µGC, shown in Fig. 7.2-7, indicates that no separation occurred during permeation from synthetic air.

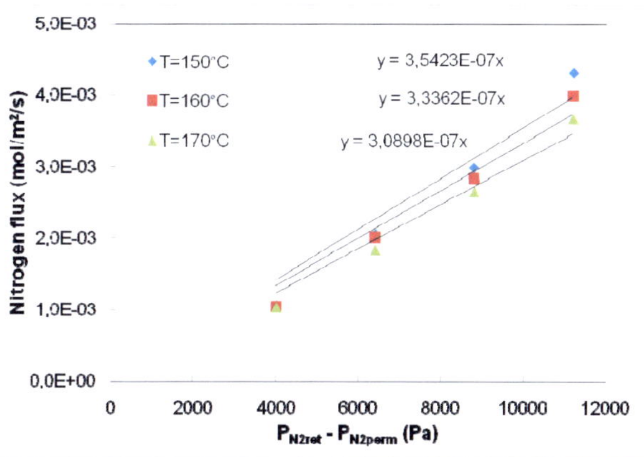

Fig.7.2-7: Permeated N_2 measured during characterization of the UF-TRUMEM membrane.

From the flow data of Fig. 7.2-7, N_2-permeances of 3.54×10^{-7} mol/m²/s, 3.34×10^{-7} mol/m²/s and 3.09×10^{-7} mol/m²/s were measured at 150°C, 160°C and 170°C respectively. This corresponds to separation factors O_2/N_2 between 1.02 and 1.01. Knudsen selectivity, theoretically, slightly favors nitrogen transport (this is calculated using the square root of the molar mass ratio of the gas molecules) at a factor of 1.07.

As illustrated in Fig. 7.2-8, when plotting the oxygen and nitrogen permeances against $1/T^{0.5}$, linear relationships are obtained. This emphasizes the presence of a Knudsen mechanism in the gas transport through the membrane

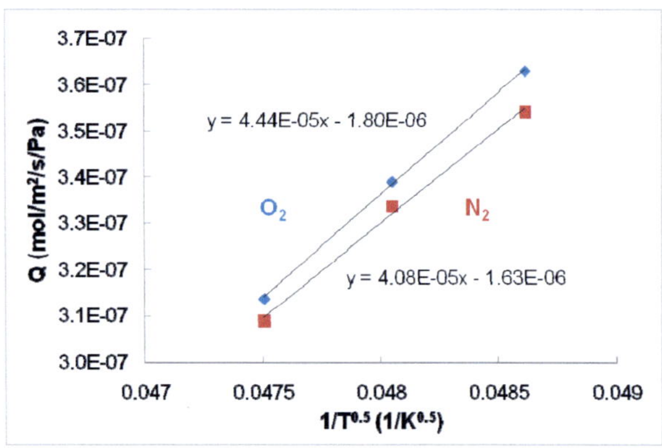

Fig. 7.2-8: Plot of the oxygen permeance of UF-TRUMEM membrane vs the Knudsen temperature dependence $1/T^{0.5}$.

8. Reactor characterization via residence time distribution measurements and in-situ gas sample extraction

8.1. Problem statement

It is important to know the flow pattern (laminar, turbulent, short cuts, dead zones) and influence of the diffusive mixing inside the reactor in order to be able to analyze the experimental data obtained from the reactor in terms of the reaction kinetics and permeance, as well as predict the reactor performance.

The reactor has 26 parallel flow channels, which are 120 mm long, 1 mm wide and 2 mm high. A 5 mm long rectangular area in front of the channels distributes the flow to all channels, and a rectangular area of the same length behind the channels guides the flow from the individual channels to the outlet bore. The bottom wall of the channels is formed by the oxygen distribution membranes, while the top wall is formed by the catalyst-coated PdCu membrane. Both membranes extend over the channels as well as the rectangular areas to the front and rear.

All channels are empty, but interrupted, every 10 mm, by ligaments in order to stabilize the microstructured plate (see Fig. 8.1-1).

One question that arises in this system is whether concentration gradients exist for the height of the channels from the oxygen distribution side to the catalyst-coated wall or whether diffusion and local deviations of the flow velocity vectors from the longitudinal axis of the channel would eliminate such gradients and effectively lead to plug flow behavior. A second concern is whether or not equal portions of the total flow are distributed to each channel. Even if plug flow through the individual channels can be assumed, a departure from the equipartition of the flow to the individual channels would still produce a broadening of the residence time distribution; it would be expected that this would affect the interpretation of the data from permeation or reaction experiments.

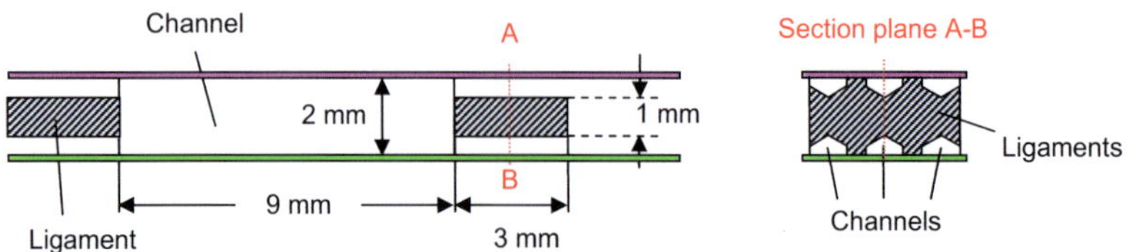

Fig. 8.1-1. Longitudinal cut through one channel on the reaction plate. Details of the geometry at the ligaments.

Fig. 8.1-2 shows the microstructured plate before assembly with the rectangular areas in front of and behind the channels. The flow to the channels depends on the pressure field, which develops in the reactor. During the reactor development, the multi-channel plate was designed with expansion and collection zones in order to facilitate the distribution of the flow in all channels.

The reactor was used, in this configuration, to perform benzene direct hydroxylation under different conditions. By assuming a plug flow distribution, the gas velocity inside channels at the extremities of the plate close to the reactor wall is considered to be the same as inside the center channels; the reaction atmosphere is thus assumed to be the same in all channels.

Fig. 8.1-2: Microstructured plate with 26 channels placed in the half reactor prior to the residence time distribution measurements

8.2. Approach

These points were mainly addressed mainly via residence time distribution measurements and 2D flow simulations using COMSOL Multiphysics. In addition, some permeation experiments were performed, which involved the extraction of gas samples from different channels, at various axial positions, via thin stainless steel capillaries and the detection of the gas concentrations.

8.2.1. Correlation of RTD measurements with reactor behavior during hydroxylation experiments

The design and utilization of the new reactor to perform benzene hydroxylation necessitates the characterization of the reactor behavior in terms of residence time distribution, in order to determine whether the behavior is ideal or non-ideal. The analysis of the distribution of the gas flow in one reaction channel compared to the flow in all channels allows for a conclusion to be drawn on the obtained type of reactor behavior.

A non-ideal behavior of the reactor caused by a non-uniform distribution of the flow to the channels or in one channel would lead to different reaction conditions inside the channels and impede the analysis of the reaction kinetics.

An ideal behavior of the reactor, with a plug flow in the reaction channels, allows for the modeling of the course of the reactions, involved in the system, along the axis of a single channel; this modeling is realized via, for example, the programming tool Matlab 2007b. The simplified system is described via a 1D pseudo-homogeneous model. The differential equation system of the model is solved, by knowing the initial conditions at x0, and gives the concentration profiles along the reactor axis x. Fig. 8.2-1 sketches a plug flow behavior, which is acting in all reaction channels. The obtained velocity profiles have no perpendicular component to the flow direction x. In this ideal case, the educt and product concentration only varies in the x direction.

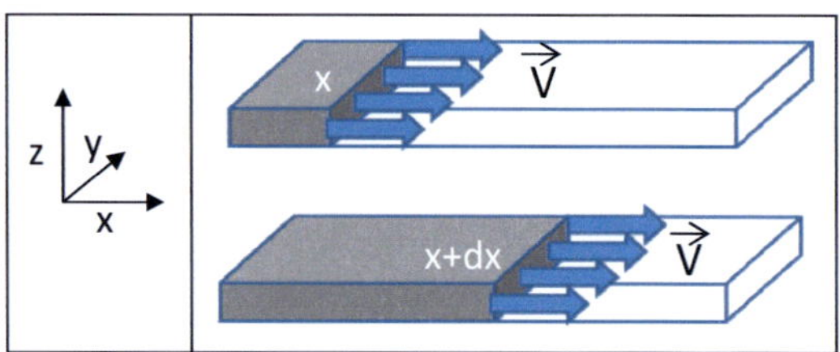

Fig. 8.2-1: Plug flow model in flat reactor type, the velocity profile is constant over the whole cross section of the reaction area, which is composed of a network of parallel channels

A rapid appraisal of whether the flow behaves ideally in one channel (no concentration gradient perpendicular to the flow direction) can be realized by comparing the characteristic time for the diffusion across the channel with the average residence time of the flow in the channel. The average residence time of the flow is defined by:

$$\tau_a = \frac{V}{F}$$

Eq (8.2-1)

where τ_a is the average residence time in s, V the channel volume in mL and F the gas flow inside the channel in mL/s.

The characteristic for diffusion across the channel is:

$$t_c = \frac{R^2}{D}$$

Eq (8.2-2)

with t_c the characteristic diffusion time across the channel in s, R the characteristic dimension of the channel in m, here the channel height, and D the diffusion coefficient of the gas in m^2/s.

The channel geometry is 120 mm x 1 mm x 2 mm (volume of 0.24 mL). Entrance flows from 10 mL/min to 100 mL/min were considered in order to be able to cover the wide range of flow rates, which are used in the double-membrane dosage or co-feed mode experiments. This results in a unitary flow through each of the 26 channels, which ranges from 0.38 mL/min to 3.8

mL/min. The average residence time τ_a in 1 channel thus ranges from 37.4 s at F=10 mL/min to 3.7 s at F=100 mL/min.

The channel height is 2 mm (characteristic dimension for diffusion) and the diffusion coefficient of oxygen in hydrogen at the reaction temperature is about 3×10^{-5} m^2/s, leading to a characteristic diffusion time across the channel of 0.1 s. τ_a in 1 channel is much higher than t_c for the whole range of entrance flows used. With τ_a much superior to t_c, one can expect that diffusion will flatten the concentration profiles across the channel and lead to plug-flow behavior of one channel. An ideal behavior of one channel is predicted. The Reynolds number characterizing the flow regime in one channel can also be estimated in order to substantiate this claim. The dimensionless Reynolds number is described as follows:

$$\mathrm{Re} = \frac{VL}{v} \qquad\qquad \text{Eq. (8.2-3)}$$

where V is the mean gas velocity (m/s), L a characteristic dimension, usually the gas travel length (m) and v is the kinematic viscosity of the gas (m^2/s), which is expressed by the ratio of the dynamic viscosity of the gas μ (kg/m.s) to the density of the gas ρ (kg/m^3): $v=\mu/\rho$. The linear velocity of hydrogen through the channel lies between 8.3×10^{-2} m/s and 8.3×10^{-1} m/s where a channel cross-section of 1 mm x 2 mm and entrance flows ranging from 10 mL/min to 100 mL/min are taken as the basis. With a channel height of 2 mm, $\mu_{H2}=8.74 \times 10^{-6}$ kg/m.s at 20°C and $\rho_{H2}=9.0 \times 10^{-2}$ kg/m^3, the Reynolds number in a channel varies from 1.7 at F=10 mL/min and 17 at F=100 mL/min. The flow regime in a single channel is laminar for all the tested entrance flows due to the fact that the critical Reynolds number, for entering the turbulent flow regime, lies at 2300 for a tubular continuous reactor. The laminar flow in one channel (maximal flow at the center of the channel and null flow at the wall) is flattened by the rapid diffusion of the gas across the channel so that no concentration gradient appears in the perpendicular flow direction.

In order to study the residence time distribution of a gas phase in a continuous reactor, a constant tracer concentration is introduced at the reactor entrance at the time defined as t=0. A sensor connected to the reactor end detects and measures the tracer concentration over time. The tracer concentration $c(t)$ gradually increases until it reaches the initial concentration c_0. The step function F(t) normalized by c_0 is:

$$F(t) = \frac{c(t)}{c_0}$$

Eq. (8.2-4)

F(t) follows the evolution of the tracer concentration at the reactor end from 0 to 1 in absolute value. At F(t)=1, only tracer molecules are found in the reactor; the tracer has "pushed" all molecules present, before its introduction in the reactor, out of the reactor. In a continuous reactor, the residence time distribution curve E(t) is obtained by deriving the function F(t):

$$E(t) = \frac{dF}{dt}$$

Eq. (8.2-5)

Fig. 8.2-1 sketches the theoretical signals measured for a step function F(t) and its related residence time distribution E(t) in the case of a continuous tubular reactor.

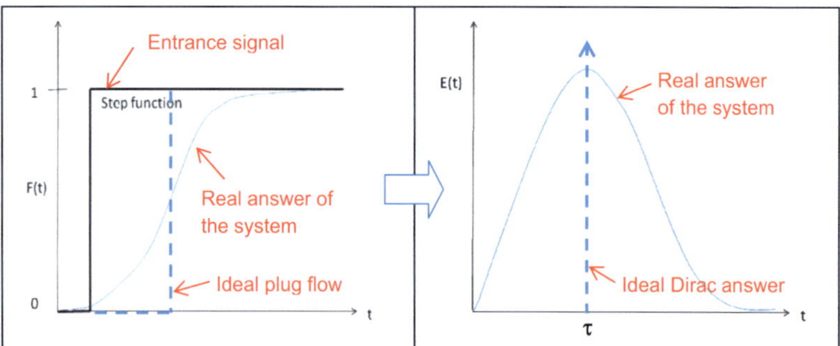

Fig. 8.2-1: Theoretical signal obtained for a step function F(t) and related residence time distribution curve E(t) in a continuous tubular reactor

In order to measure the residence time of gas molecules in the different plate configurations described above, the reactor was coupled with a hydrogen sensor for measurements of concentration variations in a short period of time. The experimental procedure employed is described below.

The E(t) values were derived from the experimental F(t) profile by taking the result obtained, when the central difference was divided by the time interval; this is described in Eq. (8.2-6).

$$E(t_i) = \frac{F(t_{i+1}) - F(t_{i-1})}{t_{i+1} - t_{i+1}}$$

Eq. (8.2-6)

8.2.2. Description of the experimental setup for RTD measurements

The experimental setup, used to determine the residence time distribution in the reactor, is composed of a nitrogen/hydrogen switch system at the reactor entrance, with the possibility to apply different gas flows, and a hydrogen sensor directly placed at the reactor exit for the measurement of the hydrogen concentration (or partial pressure in this case) over time, as sketched in Fig. 8.2-1. The hydrogen sensor and readout device were manufactured by Orbisphere Laboratories (Neuchatel/Geneva, Switzerland). The sensor is equipped with a polymer membrane 2995A (Orbisphere), which has a response time of 6 s in the gas phase according to the manufacturer specifications. The data acquisition is realized every 10 s (maximal acquisition frequence available) and is monitored via an Orbisphere-supplied software. The membranes are not desired for the residence time measurements in the reaction channel, as the Pd-based membrane would leads to some hydrogen adsorption on its surface and would thus modify the measurements. For this reason, both membrane dosage systems have been switched off; and the membranes have been replaced by aluminum plates in order to block the membrane inlets and outlets.

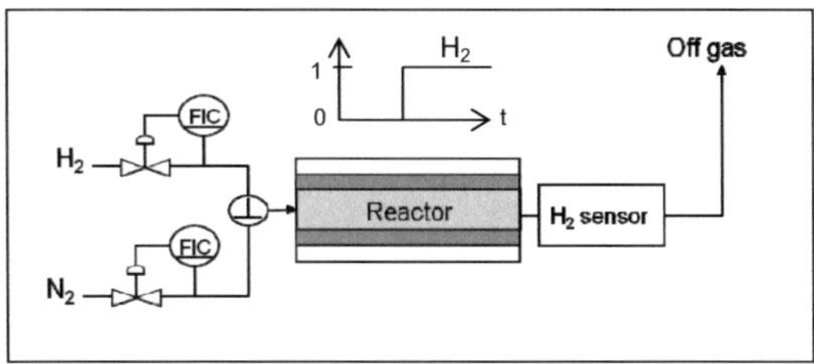

Fig. 8.2-1:   Sketch of the experimental setup for residence time distribution measurements in the reactor

The reactor was swept with nitrogen (100% N_2 in central channel) for the measurements. At a precise time t_0, the feed was switched to hydrogen (flow rate adaptable) and the increase in hydrogen partial pressure over time, measured by the sensor, was saved in the computer program. Experiments with several hydrogen flows (ranging from 8 mL/min to 32 mL/min) were carried out in order to observe the trend in the development of the shape of the residence time distribution curves. Broader distributions are expected for lower flows. After switching from nitrogen to hydrogen, it was observed that the hydrogen sensor responded at the third measurement point. The initial null values correspond with the time required for the first hydrogen molecules to travel into the reactor, together with the system response. Blank experiments without the reactor, which was replaced by a short tube between the hydrogen inlet and hydrogen sensor, led to a system response for the first measurement after 10 s (brief travel time + response time of the sensor membrane).

The setup used is shown in Fig. 8.2-2; the reactor with dead-end membrane mode has the hydrogen sensor at its output; the readout device and data acquisition system are also depicted.

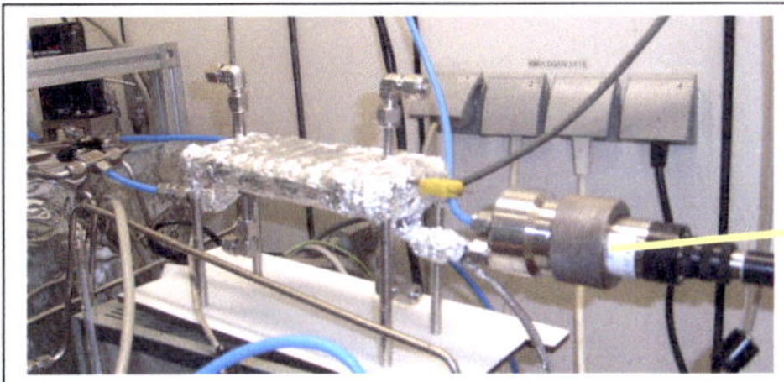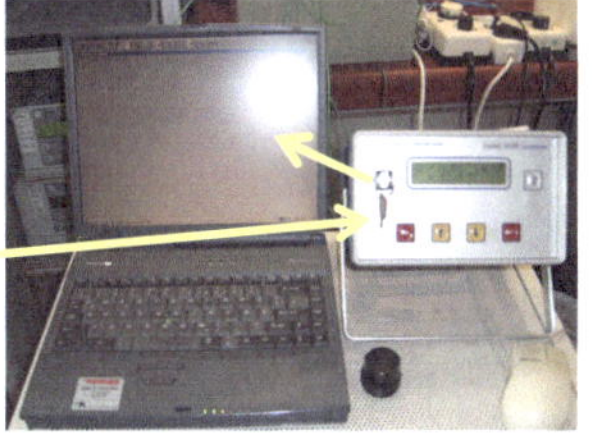

Fig. 8.2-2:   Pictures of the experimental setup with hydrogen detector at reactor output and data acquisition system

The hydrogen partial pressure in the reactor was measured as soon as the gas flow was switched from nitrogen to hydrogen. Measurements were recorded every 10 s and an end value of 1 was reached after a certain period. This resulted in the F(t) curve for different entrance flows. The residence time distribution curve E(t) was calculated from the F(t) acquired curves via data derivation following Eq. (8.2-6).

In the case of the double-membrane dosage experiments detailed later in this work, the entrance nitrogen flow was about 50 mL/min, which results in an average residence time in the reactor of 11 s, bearing in mind that the available inner volume, from the hydrogen inlet to the hydrogen sensor, is approx. 9.2 mL. With a relevant analysis domain of ca. 10 times the average residence time, one obtains about 10 measurement points on the RTD curve. If the entrance flows is 100 mL/min as for co-feed experiments, then the average residence time becomes 5.5 s and only 5 acquisition points are obtained for the plotting of the RTD curve. In this case, the used sensor, with an acquisition frequency of 10 s, is too slow to allow for an adequate study of the residence time distribution and reactor behavior. Lower flows are required in this case of such a sensor; this is why the 8 mL/min to 32 mL/min range was used here, even if it is not representative of the flows employed as part of the hydroxylation experiments. The results obtained for the RTD measurements can thus not be directly used for the characterization of the reactor under real experimental conditions. It must, however, be borne in mind that higher flows entail a higher pressure loss along the channels; this, in parallel, will improve the gas distribution inside the channels where they all have the same geometry. The RTD measurements realized here, at reduced entrance flows, represent a less convenient configuration than that which applies during permeation measurements or the hydroxylation experiments.

8.3. Residence time distribution results

8.3.1. Reactor plate with 26 channels

Fig. 8.3-1 shows the F(t) and E(t) curves for the plate with 26 gas channels at different gas flows used, ranging from 8 mL/min to 32 mL/min.

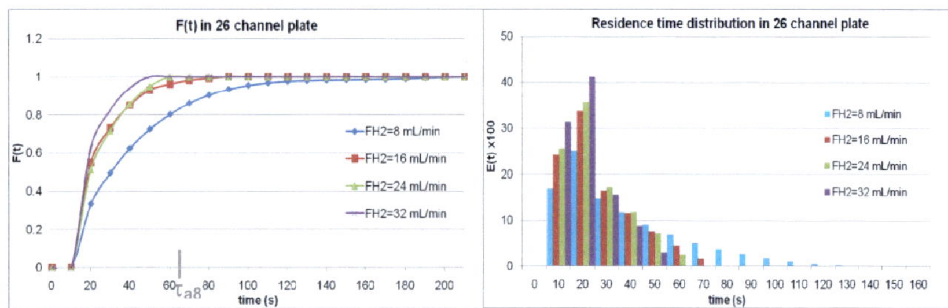

Fig. 8.3-1: F(t) curve for hydrogen crossing the reactor via the 26 channel plate, and residence time distribution curve with the 26 channel plate.

The residence time distribution curves logically feature a narrower distribution as the hydrogen flow increases. Lower flows lead to stretched distributions. One can see that all the distribution curves appear quite wide; this could be due to flow maldistribution through the 26 channels for all the tested entrance rates. This tends to indicate that the flow behavior in the reactor is quite far away from plug flow, due to an inhomogeneous flow distribution in the channels (a same flow in all parallel channels theoretically leads to a narrow distribution of residence time). There could also be an effect of the areas in front and behind the channel plate, which broaden the distribution, as their volumes also participate in the reactor flow.

A quantification of the data collected above can be realized based on the mean residence time, which applies for the different used flows.

Using discrete experimental data, the mean residence time τ of the gas molecules can be estimated via the calculation of the central moments:

$$\mu_n = \sum_{i=1}^{N} \bar{t}^n . E_i . \Delta t_i \qquad \text{Eq. (8.3-1)}$$

where μ_n is the central moment of order n, $\bar{t}$ is the middle of the discrete interval considered, Δt_i the width of the interval, E_i its height and N the number of intervals taken from the experimental curves. The E(t) diagrams are shown in the form of segments and not points (1 segment per experimental point), with Δt_i being equal to the acquisition frequency (10 s), for practical reasons in the experimental determination of the discrete intervals from the collected data.

The mean residence time τ is then:

$$\tau = \frac{\mu_1}{\mu_0} \qquad \text{Eq. (8.3-2)}$$

and the square of the variance σ, describing the dispersion around the mean value is:

$$\sigma^2 = \frac{\mu_2}{\mu_0} - \tau^2 \qquad \text{Eq. (8.3-3)}$$

Table 8.3-1 summarizes the calculation results for the mean residence time for the flow through the reactor and the variance for the 4 different used entrance flows. The calculations were done using the segment partition as shown in Fig. 8.3-1.

Table 8.3-1: Estimation of the mean residence time of gas molecules in the reactor for different flows through the 26 channel-plates and corresponding accessible volume

Number of gas channels on central plate	Hydrogen flow F_{H2} in mL/min	Mean residence time τ calculated via Eq. (8.3-2) (s)	Variance σ (s) around τ via Eq. (8.3-3)	Mean residence time τ_a expected via Eq. (8.2-1) (s)
26	8	42	25	69
26	16	31	15	34
26	24	29	13	23
26	32	26	10	17

The mean residence time experimentally measured varies from 42 s at 8 mL/min down to 26 s at 32 mL/min. For an inner accessible volume of 9.2 mL from hydrogen inlet to sensor, one expects a mean residence time of 69 s at 8 mL/min, 34 s at 16 mL/min, 23 s at 24 mL/min and finally 17 s at 32 mL/min. These values differ from those measured experimentally, with the scattering in a smaller range (42 s to 26 s). The measured residence time is shorter than expected in the case of low entrance flows. A possible explanation for this would be the presence of dead zones in the reactor (possibly at the edges of the distribution and collection zones), which are not accessible to the gas. The accessed effective reactor volume would thus be smaller than that calculated from the reactor inner geometry. The measured mean residence time was, on the other hand, longer than that expected in the case of higher entrance flows, up to 32 mL/min (configuration closer to that which was experimentally applied). This behavior could be due to an inaccurate calculation of the reactor inner volume, which is accessible to the gas. The inner volume is probably higher than the estimated 9.2 mL (complex channel plate geometry with interrupted ligaments).

Other plate systems were manufactured in order to investigate the role of the volumes generated by the distribution and collection areas inside the reactor and used for the determination of the RTD curves in the reactor. Plates with only 1 channel and 3 channels, respectively, were produced in order to study the broadening effect due to the distribution of the flow to the channels. A narrower RTD curve would be obtained through 1 channel in comparison to 3 or 26 channels for the mean residence time. The previous results of Table 8.3-1 will be further detailed and discussed as they will be compared with the results obtained through 1 and 3 channels.

8.3.2. Reactor plate with 1 and 3 channel(s)

Two additional plates with one and three channels are showed in Fig. 8.3-2. The channels have the same geometry as for the 26 channel-plate.

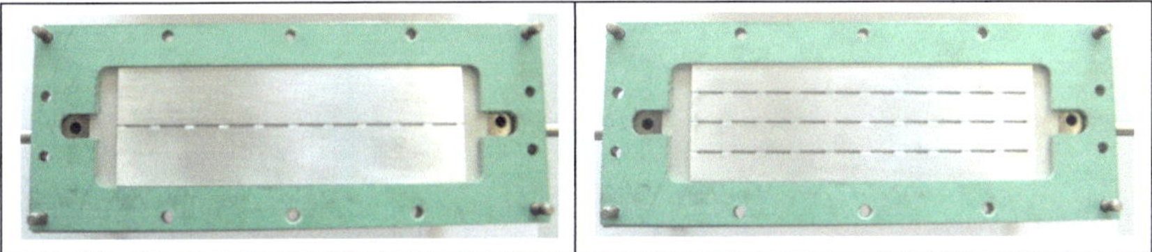

Fig. 8.3-2: Microstructured plates with 1 and 3 channel(s) placed in the half reactor before RTD measurements.

The same procedure, using the hydrogen sensor with 4 different entrance flows, was used to perform the residence time distribution measurements with the 2 new channel plates. The F(t) and E(t) curves, obtained from 1 channel and 3 channels, can be seen in Fig. 8.3-3 and Fig. 8.3-4 respectively.

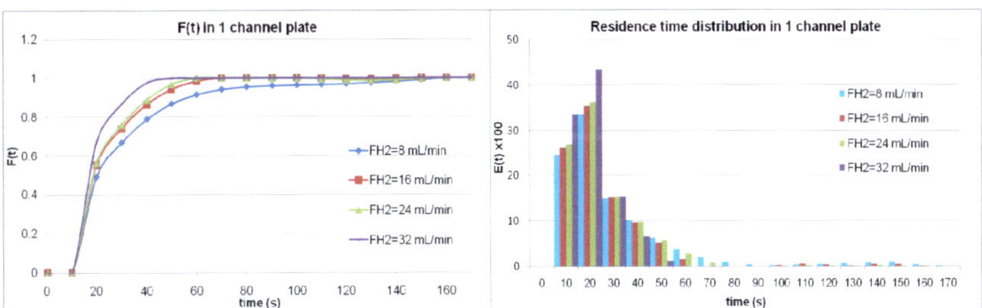

Fig. 8.3-3: F(t) curve for hydrogen crossing the reactor via the single channel plate, and residence time distribution curve with the single channel plate.

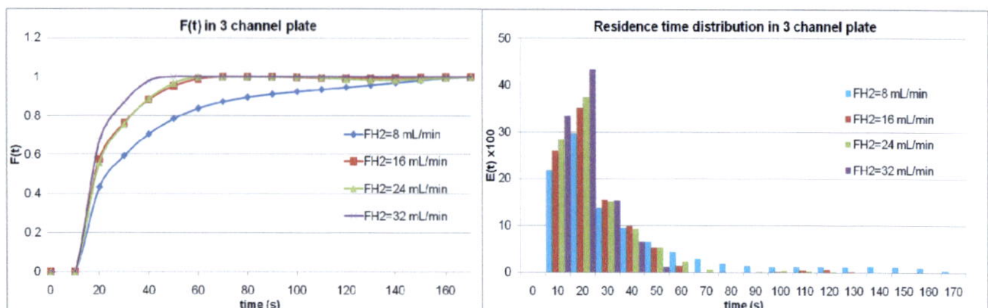

Fig. 8.3-4: F(t) curve for hydrogen crossing the reactor via the 3 channel plate, and residence time distribution curve with the 3 channel plate.

The mean residence time and the variance values calculated from the residence distribution curves of Figs. 8.3-3, 8.3-4 and also 8.3-1 are summarized in Table 8.3-2. The expected mean residence time calculated using Eq. (8.2-1) is also given in Table 8.3-2. This time, smaller inner volumes were used in the calculation due to the limited number of channels in the plates (i.e. 3.2 mL with the 1 channel plate and 3.7 mL with the 3 channel plate).

Table 8.3-2: Estimation of the mean residence time of hydrogen molecules in the reactor for different flows and channel systems (1, 3 and 26 channel-plates) and corresponding accessible volume

Number of gas channels on central plate	Hydrogen flow F_{H2} in mL/min	Mean residence time τ calculated via Eq. (8.3-2) (s)	Variance σ (s) around τ via Eq. (8.3-3)	Mean residence time τ_a expected via Eq. (8.2-1) (s)
1	8	35	31	24
1	16	31	20	12
1	24	30	17	8
1	32	25	9	6
3	8	42	34	28
3	16	30	16	14
3	24	29	15	9
3	32	25	9	7
26	8	42	25	69
26	16	31	15	34
26	24	29	13	23
26	32	26	10	17

In the single channel configuration, the mean residence time values vary from 35 s for a flow of 8 mL/min down to 25 s for a higher entrance flow of 32 mL/min. At the same time, the variance around τ decreases from 31 s down to 9 s, indicating a narrowing of the distribution as the flow increases. The observed trend again fits with the expectations, as higher entrance flows result in shorter mean residence times and narrower distributions. In the case of a single channel plate, the RTD curve is, however, expected to be narrower than that through a 26-channel plate. Here the RTD curve is even wider (higher sigma value) than through the 26 channels, even though the mean residence time is shorter due to smaller reactor volume (35 s in 1 channel in comparison to 42 s in 26 channels at F=8 mL/min). This underlines the significant influence of the distribution and collection volumes, shown in Fig. 8.2-5, on the reactor behavior (flow distributor effect in a channel network or widening of the gas path with 1 channel). Note that the mean residence time values, at F=16 mL/min, 24 mL/min and 32 mL/min respectively, are quasi-identical regardless of whether the 1-channel or 26-channel plate is used; this indicates a probable control of the mean residence time of the gas flow in the reactor by the distribution and collection zones.

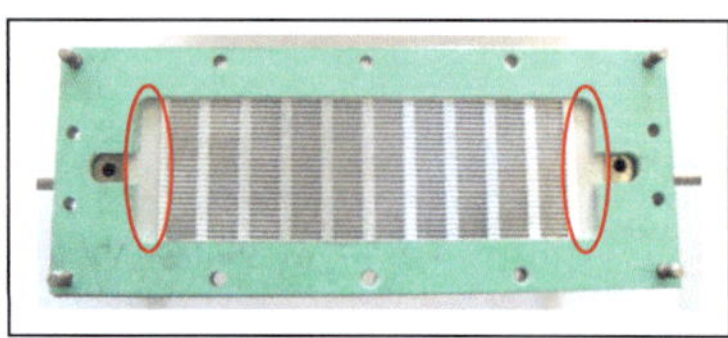

Fig. 8.3-5: Detailed view of the gas distribution and collection areas at the reactor entrance and exit

The observations made for the 1-channel plate also apply for the 3-channel plate. In terms of the mean residence time, the 3-channel plates show similar results, except for the first value (τ=42 s at F=8 mL/min for the 3- and 26-channel plates as opposed to τ=35 s in the 1-channel plate). A higher imprecision (broader RTD) is expected with the 1- and 3-channel plates, which exhibit a smaller inner volume, particularly at higher flows due to the used sensor (data acquisition only every 10 s). Smaller volumes shorten the expected mean residence time. The front part of the RTD can thus not be measured due to the low data acquisition frequency. As the first measurement values are taken after 10 s, a large part of the distribution curve cannot be seen in the case of high entrance rates (the expected mean residence time values are below 10 s). The use of a different type of sensor is required in order to enable measurements over the whole range of involved entrance rates. Wider RTDs are also observed when the system variance is compared, as both the 1- and 3-channel plates exhibit higher variance values in comparison to the 26-channel plate.

Using Figs. 8.3-1, 8.3-3 and 8.3-4, it was possible to plot the residence distribution curves of the 3 different plate configurations, together on the same graph, for a given entrance flow. This was realized for the 24 mL/ min and 32 mL/min hydrogen flows respectively, which are closer in terms of their range to that used during benzene hydroxylation in the experiments. The distribution curves are showed in Fig. 8.3-6.

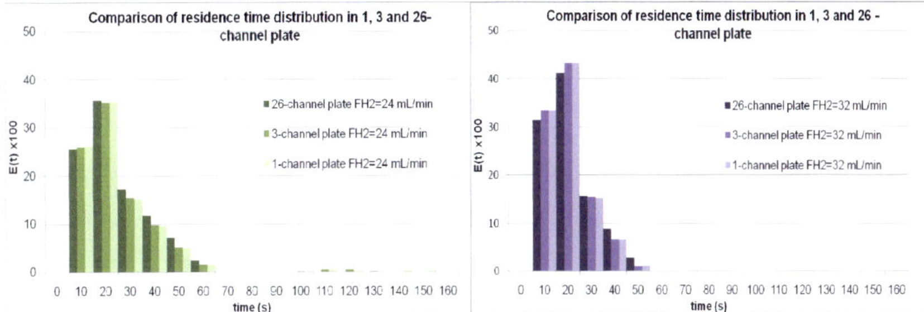

Fig. 8.3-6: Comparison of the residence time distribution curves obtained at 24 mL/min and 32 mL/min of hydrogen flow for the 3 different plate structures.

The curves of the 3 plate configurations are, in both cases, superimposed. They have the same shape, sometimes the exact same amplitude. A narrower distribution was expected with the single-channel plate, due to the absence of a distribution into the different parallel channels, in comparison to the other systems. This was clearly not the case. There is no distinction between the 3 plate systems in terms of the measured RTD. This clearly indicates that the distribution/collection areas in the reactor control the RTD of the reactor.

Conclusion on the experimental RTD measurements:

Two main observations can be derived from the experimental reactor characterization. After taking into account the convective and diffusive flows acting in one channel (absence of concentration gradient due to the rapid gas phase diffusion, which flattens the profiles), it can be expected that each channel itself exhibits plug flow behavior. Experimental residence-time distribution measurements resulted in rather broad distributions even for increased entrance flows, up to 32 mL/min, which approximate the experimental conditions employed in the reactor. This indicates non-plug flow behavior in the multi-channel system, due to the observed strong influence of the distribution and collection areas on the overall flow behavior in the reactor. A non-plug flow behavior of the reactor could be caused by a non-homogeneous distribution of the flow to the different channels. An additional simulation step is required in order to obtain more information on the behavior of the flow in the channels.

8.4. 2D simulation with COMSOL Multiphysics

A model of the gas streaming into the reactor channels was developed using the software COMSOL Multiphysics. The aim was to bypass the difficulties, which appeared in relation to the reactor configuration for the experimental determination of the RTD through several types of channel plates, by simulating whether an equidistribution of the flow into the channels actually applies under the employed experimental conditions. A convective-diffusive model based on the Navier-Stokes equations was selected for the simulation.

The Navier-Stokes equations are non-linear second order differential equations, which allow for the description of a fluid motion in a particular structure (tube, reactor, etc.), assuming that the fluid stress is the sum of a diffusive viscous term (proportional to the gradient of velocity) plus a pressure term.

The Navier-Stokes equations, used in the COMSOL software, are described in Eq (8.4-1) and Eq. (8.4-2). The incompressibility condition applies in the model (constant fluid density).

$$\rho\left(\frac{\partial u}{\partial t}+u\frac{\partial u}{\partial x}+v\frac{\partial u}{\partial y}+w\frac{\partial u}{\partial z}\right)=-\frac{\partial p}{\partial x}+\frac{\partial}{\partial x}\left[2\mu\frac{\partial u}{\partial x}\right]+\frac{\partial}{\partial y}\left[\mu\left(\frac{\partial u}{\partial y}+\frac{\partial v}{\partial x}\right)\right]+\frac{\partial}{\partial z}\left[\mu\left(\frac{\partial w}{\partial x}+\frac{\partial u}{\partial z}\right)\right]$$

$$\rho\left(\frac{\partial v}{\partial t}+u\frac{\partial v}{\partial x}+v\frac{\partial v}{\partial y}+w\frac{\partial v}{\partial z}\right)=-\frac{\partial p}{\partial y}+\frac{\partial}{\partial y}\left[2\mu\frac{\partial v}{\partial y}\right]+\frac{\partial}{\partial z}\left[\mu\left(\frac{\partial v}{\partial z}+\frac{\partial w}{\partial y}\right)\right]+\frac{\partial}{\partial x}\left[\mu\left(\frac{\partial u}{\partial y}+\frac{\partial v}{\partial x}\right)\right]$$

$$\rho\left(\frac{\partial w}{\partial t}+u\frac{\partial w}{\partial x}+v\frac{\partial w}{\partial y}+w\frac{\partial w}{\partial z}\right)=-\frac{\partial p}{\partial z}+\frac{\partial}{\partial z}\left[2\mu\frac{\partial w}{\partial z}\right]+\frac{\partial}{\partial x}\left[\mu\left(\frac{\partial w}{\partial x}+\frac{\partial u}{\partial z}\right)\right]+\frac{\partial}{\partial y}\left[\mu\left(\frac{\partial v}{\partial z}+\frac{\partial w}{\partial y}\right)\right]$$

Convective term *(related to the fluid density and the velocity field)* **Diffusive term** *(related to the fluid viscosity, the stress tensor - obtained from the 2nd derivative of the velocity components - and the pressure field acting on the fluid)*

Eq. (8.4-1)

The continuity condition (divergence-free velocity field) states that:

$$\frac{\partial u}{\partial x}+\frac{\partial v}{\partial y}+\frac{\partial w}{\partial z}=0 \qquad\qquad \text{Eq. (8.4-2)}$$

where (u, v, w) are the 3 components of the fluid velocity field (m/s) in Cartesian coordinates (x,y,z), ρ the fluid density (kg/m^3), p the pressure field and μ the dynamic viscosity of the fluid (Pa.s).

A two dimensional drawing of the central reaction area was realized, which is composed of the microchannel network. A gas distribution zone was required in order to enable the gas flow to travel in all parallel channels of the reactor. At the channel exit, the gas lines are collected in a similar merging area before the flow exits the reactor. The drawing is at scale 1; all dimensions correspond to those of the reactor designed in the experimental part. The reaction area is shown in Fig. 8.4-1.

Fig. 8.4-1: Design of the central area of the hydroxylation reactor (drawn at scale 1) using COMSOL Multiphysics

A nitrogen stream (benzene carrier gas) inside the reactor was simulated under standard condition of temperature. The gas viscosity was μ=-7.88x10^{-12}T^2+4.427x10^{-8}T+5.204x10^{-6} Pa.s, while the density was ρ=(28.10^{-3}P)/(8.314T) kg/m^3. The pressure P at the reactor entrance was influenced by the applied gas flux, while the pressure at the reactor exit was set at 10^5 Pa.

Only the convective nitrogen flow was simulated here. It was assumed that the hydrogen and oxygen flows from the two membranes (above and below – not represented in the drawings) into the central area are homogeneous due to the membrane diffusion processes.

The following boundary conditions were applied for the model. No slip at the channel and reactor walls, the velocity vector at the wall was null. An initial gas entrance velocity of 70 mm/s was applied, which corresponds to an entrance flow of 50 mL/min (range of double-membrane dosage experiments) after the flow is divided by the reactor inlet cross-section. The gas entrance velocity induces a pressure drop inside the reactor and in the channels, which was calculated by the software. Two different mesh densities were tested. The first mesh had a rather low amount of nodes: 7303. This one was refined several times in order to obtain a higher degree of precision when calculating the velocity field. The second mesh had 90079 nodes.

8.4.1. Influence of the meshing on the simulation results

Fig. 8.4-2 shows the 2D representation of the simulation result using the finer meshing in the model and an entrance linear velocity of 70 mm/s.

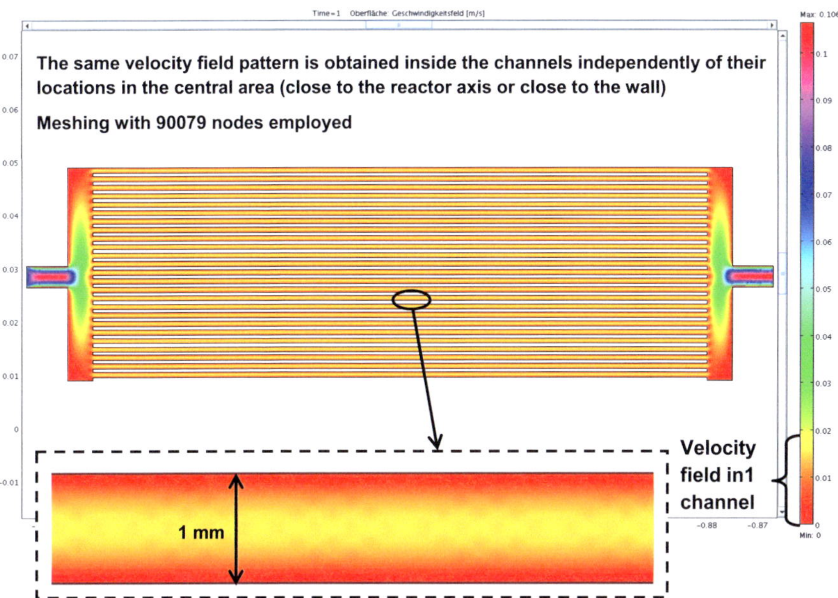

Fig. 8.4-2: u-component of the velocity field simulated in the reactor employing the usual configuration (presence of the multichannel plate for gas distribution in the central area) with the fine meshing

It can be seen from the color scale that the velocity field is similar in all channels, with an average value of 20 mm/s per channel. The maximum field velocity is obtained at the gas inlet and outlet along the reactor axis with a value of 106 mm/s. It is only the corners of the distribution area and collection zone, which are not properly accessed by the gas. The gas distribution appears homogeneous regardless of whether the channel is in the middle of the plate or far away from the reactor axis and gas inlet.

This effect is due to a similar pressure drop in all channels, as is depicted below in Fig. 8.4-3. A pressure drop of 17500 Pa was obtained in the reactor for an entrance velocity of 70 mm/s. All channels are similarly affected by the pressure drop; hence the equal distribution. The simulation of the pressure field between points A and B (reactor entrance to channel entrance), B and C (channel entrance to channel exit) and C and D (channel exit to reactor exit) showed that most of the pressure drop (16400 Pa) occurs in the channels and not in the distribution/collection zones. This ensures a homogeneous distribution of the flow to the channels.

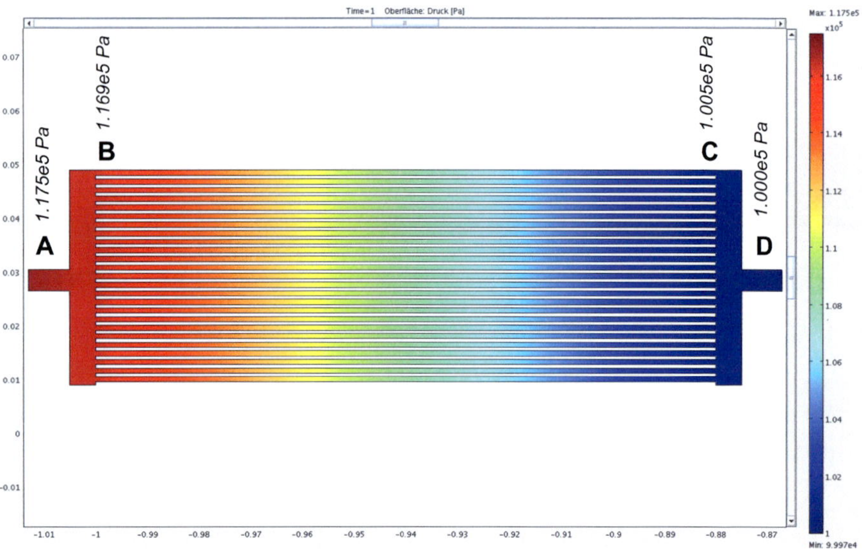

Fig. 8.4-3: Pressure drop in the reactor from entrance to exit corresponding to the velocity field simulated in Fig. 8.4-3

The simulation was run again with the same parameters (same entrance velocity and same outlet pressure) but using the 1st mesh density (7303 nodes) in order to analyze the influence of the node density on the velocity field. Due to the similarity with what was obtained in Fig. 8.4-2, the result of the simulation is not displayed.

The more limited number of nodes available for calculation purposes, in this connection, was apparent in the channels (repetition of the same geometric pattern). The results were, however, similar to those obtained with the finer meshing. The velocity field was homogeneous in all parallel channels and the pressure field was identical to that illustrated in Fig.8.4-3. The maximum velocity at the gas inlet resulted in a slightly lower value of 0.0962 m/s instead of 0.106 m/s, which may be due to the lower amount of calculation points. The simulation is thus less precise here.

Even though the mesh density of the area available to the gas does not modify the simulated velocity field, the higher mesh density is preferred as it allows for more precise calculation results.

8.4.2. Influence of the entrance flow

A simulation with a twice higher entrance velocity, i.e. 140 mm/s (eq. to a flow of 100 mL/min) instead of 70 mm/s, was realized in order to investigate possible changes in the velocity profile inside the channels. The same gas distribution pattern in the channels was observed, with a proportional increase in the field amplitude as the only noticeable change (0.211 m/s as max. flux instead of 0.106 at 70 m/s). The 2D result is not displayed here due to the similarity of the respective distributions. The same pressure drop pattern in the reactor is also observed in this case. The entrance pressure, which is related to the increased flux, is 1.349×10^5 Pa (as opposed to 1.175×10^5 at 70 mm/s); the exit pressure was kept constant at 1×10^5 Pa. In accordance with the simulation, a change in the entrance velocity, in the range of what was experimentally applied, does not influence the shape of the velocity profile in the reactor or the homogeneity of the gas distribution in the channel network.

8.4.3. Gas flux at the channel exits

The gas flux leaving the channels was independently calculated for each channel. The intensity of the gas flux exiting each channel is plotted in Fig. 8.4-4.

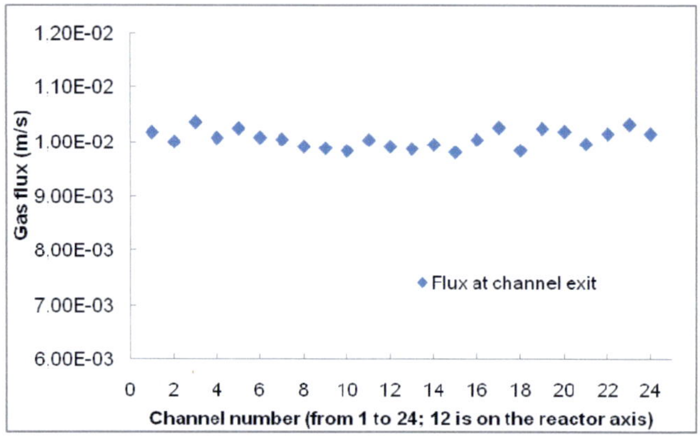

Fig. 8.4-4: Gas flux exiting the channels

The stable flux values, with an average value of 1.00×10^{-2} m/s and a standard deviation of 1.60×10^{-4} m/s, are apparent. The gas fluxes, exiting the central channels, show slightly lower values than those exiting the channels close to the reactor wall (far from the reactor axis). Even though this effect is very mild and could be due to limitations in the used model, it is still unexpected, as the gas inlet is located on the reactor axis, which faces the central and thus more easily accessibly channels. The maximal value of 1.03×10^{-2} m/s was simulated close to the reactor wall, while the minimum, 9.83×10^{-1} m/s, was obtained close to the reactor axis. The flux calculated at the reactor outlet was 2.41×10^{-1} m/s; this also corresponds with the sum of the fluxes for all channels. It is 99.2% of the flux obtained at the inlet, which shows that the material balance has been respected (no accumulation or reaction). The slight deviation of 0.8%, from a perfect balance, is in the range of the numerical precision allowed for by the simulation. According to the performed simulation, an inhomogeneous distribution of the flow in the reactor, under the experimentally applied conditions, is not expected.

The main benefit of the analysis realized using COMSOL Multiphysics comes from the observation that the gas flow is equally distributed in all parallel channels used for the benzene hydroxylation, as the flow in the channels is similar regardless of their position in the reactor (center or reactor wall). The equidistribution of the flow in all channels, regardless of their location in the reactor, which was expected when designing the multichannel plate, could at least be shown by the simulation.

Due to this effect, the gas dosage and reaction kinetics inside the reactor along the axis can be described using a simple 1D-model with Matlab, where a single channel is considered. The hydroxylation reaction will follow the same kinetics in all parallel channels and the real case can be simplified by studying the reaction kinetics in one channel, due to the gas flow being the same in all channels.

8.5. Experimental proof of concept with in-situ measurements of H_2 flow inside the reactor

A further modification of the reactor setup was carried out in order to investigate the equidistribution of the flow and allow for a measurement of the concentration profiles at different locations along the reactor axis. The modification consisted of integrating small stainless steel capillaries (inner diameter of 0.7 mm) into the sealing, which separates the 2 membranes, in order to enable the extraction of gas samples at different locations on the reactor axis. Fig. 8.5-1 shows a view from the top of the reactor in open state, with capillaries extending into the channels of the central microstructured plate for an in-situ gas sample extraction.

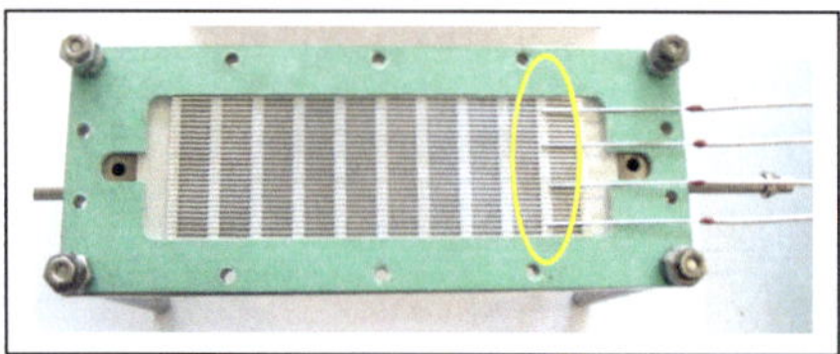

Fig. 8.5-1:   View of capillaries integrated in the reactor sealing for the extraction of gas samples inside the reaction area

Fig. 8.5-2 shows the assembled reactor with 4 capillaries integrated into the sealing, 2 close to the entrance and 2 close to the end, for gas sample extraction when the hydrogen permeation experiments are in progress. A system of taps, connected to the µGC for gas phase analysis, enables the switching from one capillary to the other.

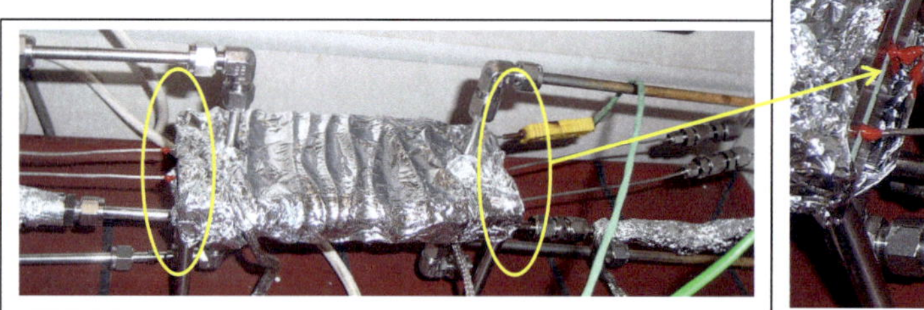

Fig. 8.5-2: View of reactor assembled with capillaries integrated in the sealing to extract gas samples along the reaction channel

Fig. 8.5-3 sketches the 4 locations of the capillaries inside the reactor where the gas samples can be taken. They are identified by the coordinates x and y according to the reactor axis, and are in both axial (positive values) and radial positions (positive and negative values).

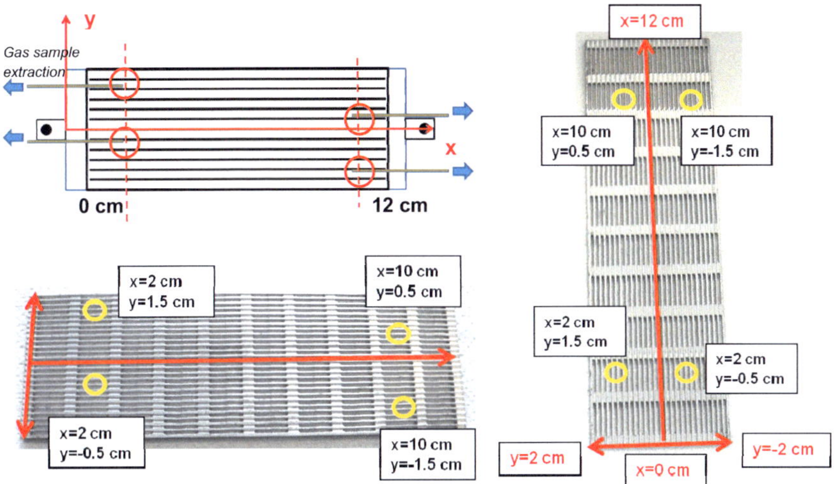

Fig. 8.5-3: Position of the capillaries inside the reaction channel plate for gas sample extraction at x=2 cm and x=10 cm along the membrane axis. Regarding their radial position along the membrane width, the capillaries allow an extraction at y=+/-0.5 cm and at y=+/-1.5 cm.

A vacuum pump model MZ 2C of the company Vacuubrand (Wertheim, Germany) was employed in addition to the µGC sampler extractor in order to compensate for the small capillary diameter. The objective was make a representative sample of the local reactor atmosphere reach the column of the µGC for compositional analysis; the effect of the aspiration may, however, affect the channels next to the capillary entrance and lead to some errors in the atmosphere composition. These errors should, however, have little effect on the analysis value. Experiments without vacuum pump were initially carried out, but the pressure drop in the channels was too significant, when compared to the drop in the capillaries, to allow for a sufficient gas sample extraction.

Permeation experiments were carried out with the 1 µm PdAu coated PdCu foil on the hydrogen side. The oxygen dosage side was obstructed by an aluminum plate in order to focus on 1 gas as the "tracer" in the multi-channel plate. No benzene hydroxylation test was performed in this configuration. The used µGC is only suitable for gas phase analysis and does not allow for a direct detection of the benzene, phenol or water, which would become trapped before reaching the column. The purpose of the system is two-fold. The first is to analyze the atmosphere composition inside the reactor as hydrogen diffuses through the membrane along the reactor axis and compare the results with the permeation laws (experiments and simulation). Secondly, an investigation of the atmosphere composition in the radial direction will help to confirm whether or an equidistribution of flow can be neasured inside the reactor channels.

A technical difficulty was encountered during sampling via the capillaries. It was noticed that the pump was not properly gas tight and that air from the atmosphere had entered the system and mixed with the gas sample during extraction. This contamination complicated the measurement of the local gas composition. By determining the amount of air (based on oxygen) and the measured dilution of the hydrogen in the sample, it was, however, possible to correct the measured composition.

Fig. 8.5-4 shows the element balance used to correct the hydrogen composition in the extracted gas sample due to an air leakage at the vacuum pump.

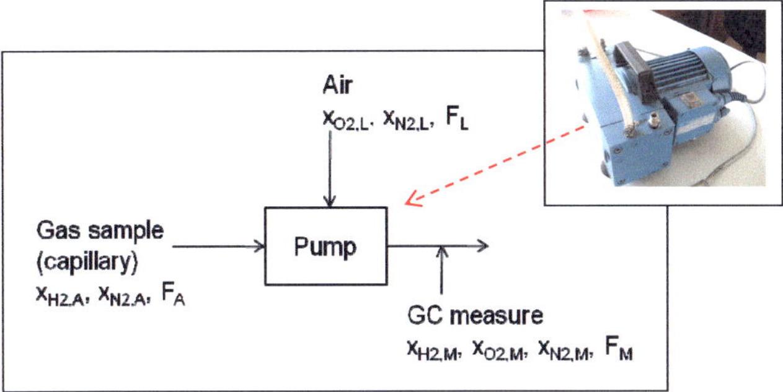

Fig. 8.5-4: Diagram of the element balance incl. the capillaries, the µGC and an air leakage at the pump used for sample extraction.

A relates to the volume fraction and flow of gas extracted at the capillaries, *L* relates to the air leakage present at the pump during sample extraction and *M* relates to the volume fraction and flow measured at the µGC (known values). The element balance of the system sketched in Fig. 8.5-4 results in:

$$\begin{cases} x_{O2,L} \cdot F_L = x_{O2,M} \cdot F_M \\[2mm] x_{N2,A} \cdot F_A + x_{N2,L} \cdot F_L = x_{N2,M} \cdot F_M \\[2mm] x_{H2,A} \cdot F_A = x_{H2,M} \cdot F_M \end{cases}$$

with $x_{O2,L} = 0.21$; $x_{N2,L} = 0.79$; $F_A = 90.2$ mL/min, $x_{N2,A} + x_{H2,A} = 1$ and $x_{H2,M} + x_{O2,M} + x_{N2,M} = 1$. $x_{H2,M}$, $x_{O2,M}$ and $x_{N2,M}$ are measured by the µGC. $x_{H2,A}$ is the unknown parameter to be determined using the equation system and the measured values.

By solving the system of equations, the values of $x_{H2,A}$ can be found via the intermediary calculation of F_A and F_M. Table 8.5-1 gives an overview of the results, in particular, with regard to the corrected hydrogen composition at the 4 locations in the reactor.

Table 8.5-1: Values of the H_2, O_2 and N_2 fractions measured (in the presence of an air leakage) and calculated (in the absence of a leakage) using the element balance of the system for samples extracted at different locations in the reactor

Position / gas fraction x_i	x=2; y=1.5	x=2; y=-0.5	x=2 average	x=10; y=0.5	x=10; y=-1.5	x=10 average
$x_{H2,M}$	0.0018	0.0016	/	0.0046	0.0042	/
$x_{O2,M}$	0.149	0.155	/	0.157	0.148	/
$x_{N2,M}$	0.849	0.843	/	0.843	0.847	/
$x_{H2,A}$	0.0062	0.0061	0.00615	0.0182	0.0143	0.01625
$x_{N2,A}$	0.9938	0.9939	/	0.9818	0.9857	/
% of x_{H2} end	28.2	27.7	27.95	82.7	65.0	73.85
F_{H2} (mL/min)	0.559	0.550	0.5545	1.64	1.29	1.465

Fig 8.5-5 shows the values of the hydrogen extracted flow according to the axial position (data from Table 8.5-1).

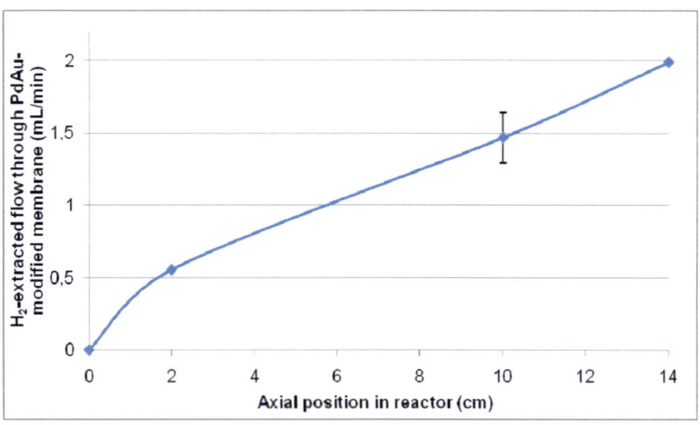

Fig. 8.5-5: Hydrogen extracted flow measured at axial position x along the reaction channels

A constant increase of the hydrogen concentration along the reactor axis is observed. This was expected due to the higher membrane surface area available for hydrogen permeation along the reactor axis. The maximal concentration was reached at the reactor end, at x=14 cm, where the whole membrane area has been used to supply permeated hydrogen.

The hydrogen permeated flows in the radial direction are illustrated in Fig. 8.5-6. In addition to the flow increase in the axial position, the behavior of the flow in the radial direction is of interest for analyzing the gas flow distribution in the different channels of the microstructured plate.

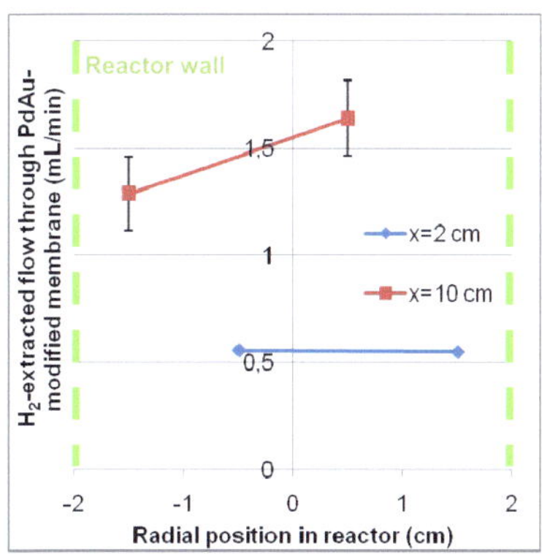

Fig. 8.5-6: Hydrogen permeated flow measured at radial position y along the reaction channels

The same flow is expected at different x for the same axial position y on the reactor axis in the case of an equal flow distribution in all parallel channels, which covers the reactor width (position ranging from y=-2 cm to y=2 cm). In Fig. 8.5-6, it can be observed that this applies for the position x=2 cm where a common flow of 0.55 mL/min was measured. For the second position along the axis, at x=10 cm, deviating hydrogen flow values were obtained at the 2 radial positions. A lower H_2 flow of 1.29 mL/min was measured in the channel close to the wall, whereas 1.64 mL/min was obtained in the channel close to the reactor axis. This represents a deviation of approx. 20% of the value of the flow along the radial axis. As only 4 capillaries were used in the modified reactor, it is not possible to state whether this radial dispersion of the flow values appears at other locations on the reactor axis. The first experimental points taken at x=2 cm indicate an equidistribution of the flow in the channels; however, the second measurements realized at x=8 cm on the axis show a 20% offset in the flow values. The deviating flows could be caused by an imperfect contact between the microstructured plate segments and the hydrogen supply membrane. This may enable radial diffusion from 1 channel to the other, although the reactor plates were solidly pressed together and a consistent driving force was applied on the membrane (approx. 3 bar of pressure at the retentate side). An equidistribution of the flow in the whole reactor is thus expected but it cannot be fully demonstrated.

9. Hydroxylation experiments in the PdCu-membrane reactor: reaction results

This chapter describes the results of the direct hydroxylation experiments carried out with different types of Pd-based catalytic surfaces and two types of membranes for the introduction of oxygen into the reactor. A large range of process parameters was employed for this purpose, e.g. the reaction temperature, partial pressure amplitude in the system and oxygen introduction mode (double-membrane dosage or co-feed).

The first group of experiments was performed with the aim of, on the one hand, evaluating the performance of two types of PdCu surface, depending on the preparation process (electroless plating, cold rolling), and on the other hand of assessing the parameter range, which allowed for the best phenol selectivity depending on the respective hydroxylation conditions. After this step, a series of experiments was carried out using several catalytic systems (PdGa, PdAu, V_2O_5/PdAu, which were deposited on the PdCu membrane with the objective of improving the phenol selectivity and reducing the effect of the side-reactions in the reactor. One of the stated goals of the surface modification was to investigate the influence of the catalysis on the gas phase phenol production from benzene.

Blind experiments without PdCu membrane were carried out at the start in order to investigate a possible catalytic effect of the steel reactor wall or silver membrane. These experiments were realized with a mixture of H_2 and O_2 and then with a mixture of H_2, O_2 and benzene with N_2. No product formation was detected during this initial step, which indicates that the surface of the Pd-based membrane was the only active surface in the investigated gas phase reactions. Changes had to be realized on the oxygen dosing side as the initially employed silver foils featured defects, which led to nitrogen diffusion into the reaction channels. The temperature was varied in the range between 150°C and 250°C. The H_2/O_2 concentration ratio in the reactor, which was calculated using the partial pressure ratio of the 2 gases measured at the reactor end, was changed by keeping the oxygen partial pressure constant and varying the hydrogen partial pressure; this was realized by adjusting the hydrogen retentate pressure. A more detailed investigation of the influence of the process parameters on the reaction conditions was also carried out by using the catalytic surface, which featured the highest product rates.

9.1. PdCu as catalytic surface

9.1.1. Reaction products

The first hydroxylation experiments with the reactor were carried out using a 15 µm thick Ag foil for O_2 supply and a 50 µm thick $Pd_{60}Cu_{40}$ foil as the H_2 selective membrane. The defect-free structure of the PdCu foil ensured the supply of dissociated hydrogen molecules during the bulk diffusion; these are required for the formation of active surface radicals. Said radicals are, in turn, involved in phenol production according to the original concept [24]. An untreated foil was initially used; it subsequently quickly became apparent that the surface of the laminated foil was inactive towards benzene conversion. No phenol was formed and only some traces of CO_2 and water were detected. Other experiments were realized using activated foils. Pd nanoparticles were deposited on the membrane surface via the thermal decomposition of Pd-II-acetate in order to increase the catalytic activity. A weight increase of 20 mg was measured after the Pd seeding process.

Phenol was subsequently detected in the system together with CO_2 and water. The amounts obtained, however, showed that CO_2 and water were the dominating products. No other products apart from phenol and CO_2 were detected (absence of hydrogenation product). Note that due to limitations in the initial experimental setup, the amount of water formed could not be determined in this series of experiments and can therefore not be discussed in this section. An improvement in the cooling trap container design made it possible, in the next series of experiments to establish the residual amount of non-organic liquid phase (i.e. water from direct hydrogen oxidation over Pd).

9.1.2. Parameter influence

The results, showing the influence of several process parameters on benzene hydroxylation, are presented in this section. These results were obtained after carrying out 9 experiments using the double-membrane reactor. Each experiment is solely focused on one process parameter.

9.1.2-1 H_2/O_2 concentration ratio

A series of experiments was carried out at 150°C using the following membrane foils: 50 μm $Pd_{60}Cu_{40}$ and 15 μm Ag. A different H_2/O_2 concentration ratio, measured at the reactor end, was applied in the reactor, for each experiment, in order to determine its influence on the phenol selectivity. The H_2/O_2 ratio was adjusted by keeping one of the gas partial pressures constant, while changing the other one. In this case, the hydrogen partial pressure was varied, while the oxygen pressure was kept constant at 0.1 bar.

Fig. 9.1-1 illustrates the phenol rate obtained with respect to the H_2/O_2 ratio. This ratio was calculated based on the amount of permeated hydrogen and oxygen detected at the reactor exit. The double-membrane dosage should result in a quite constant H_2/O_2 ratio all along the reactor axis. This can be investigated by simulating the membrane dosage and reaction process along the reactor axis. This shall be discussed in part 11

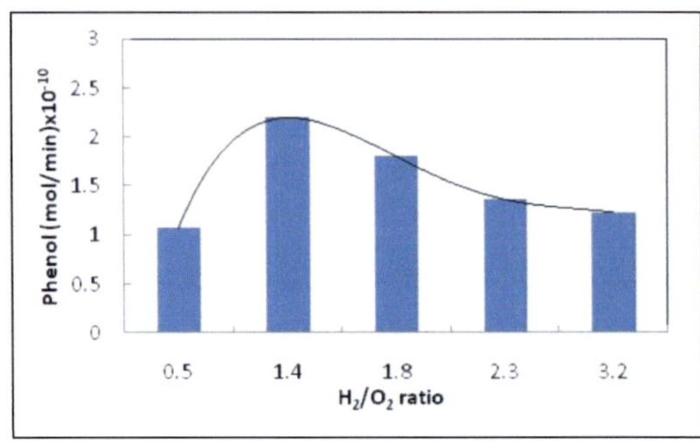

Fig. 9.1-1: Phenol production rate with respect to the H_2/O_2 concentration ratio in the reactor exit

The phenol production rate passes by a maximum for a ratio of 1.4, which corresponds to a suitable composition for benzene hydroxylation. Beyond this maximum, oxidative (low H_2/O_2 ratio) or hydrogenative (low H_2/O_2 ratio) atmospheres in the reactor increase the influence of the side-reactions. This phenomenon correlates with a similar observation from Sato et al. [26], which postulated that an adequate hydrogen and oxygen mixture favors benzene hydroxylation to phenol.

Fig. 9.1-2 shows the amount of CO_2 produced with respect to the H_2/O_2 concentration ratio.

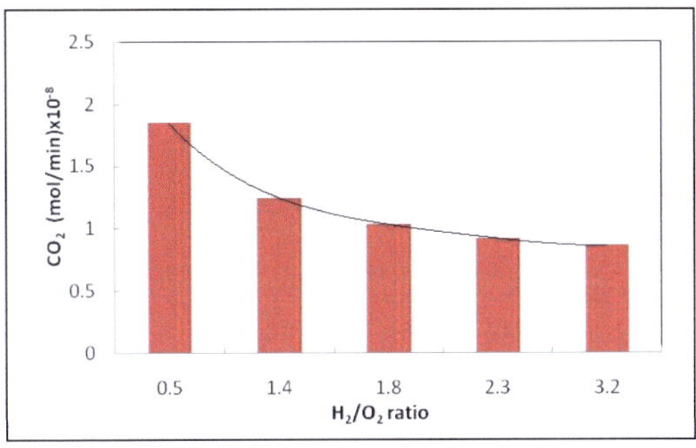

Fig. 9.1-2: CO$_2$ production rate with respect to the H$_2$/O$_2$ concentration ratio in the reactor exit

The highest amount of carbon dioxide is obtained under the most oxidative conditions, while the CO_2 production rate monotonically decreases with the increasing H_2/O_2 ratio; this is influenced by the constantly decreasing oxygen partial pressure in the reactor. Carbon dioxide remains, however, the dominating product.

While a direct comparison is not possible, similarly shaped profiles for phenol and CO_2 were reported by Sato et al. during the direct hydroxylation of methyl benzoate to methyl salicylate in their tubular membrane reactor (see Fig. 9.1-3). In this case, phenol and CO_2 are by-products generated by hydroxylation of benzene (oxidation product of methyl benzoate) and the total oxidation of methyl benzoate [26]. In the Sato case, the H_2/O_2 concentration ratio was measured along the reactor axis during a single experiment as reported by the authors (no detail is supplied on the method employed for the in-situ measurement), whereas we obtain them for different experiments performed at different H_2/O_2 ratios.

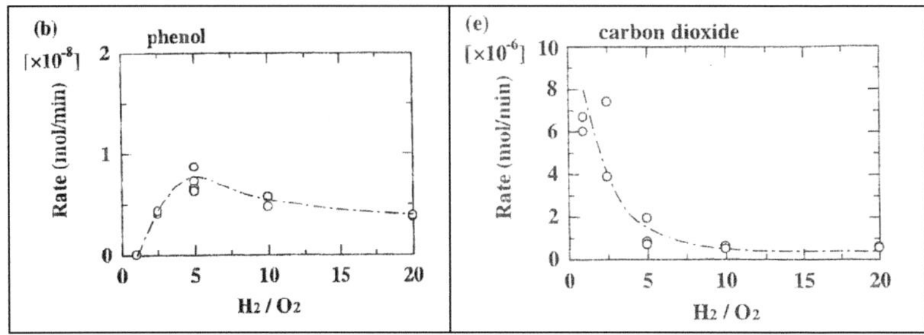

Fig. 9.1-3: Phenol (b) and CO$_2$ (e) production profiles with respect to the H$_2$/O$_2$ concentration ratio in the reactor in the case of by-products formation during gas phase hydroxylation of methyl benzoate, as observed by Sato et al. [26]

As suggested in previous papers [26, 27], the atmosphere composition has an influence on the reaction performance. The highest phenol selectivity at 9.6% was obtained for a slight hydrogen excess (H$_2$/O$_2$ ratio of 1.4) in the system. The phenol selectivity gradually decreased as the hydrogen excess increased. It was found to be quite low under oxidative conditions (H$_2$/O$_2$ ratio of 0.5) and a large amount of CO$_2$ was generated.

Fig. 9.1-4 depicts the change in the benzene conversion measured at 150°C with different H$_2$/O$_2$ concentration ratios. We logically observe a decrease as the ratio increases, a lower oxygen partial pressure leads to less benzene oxidation, and thus to less conversion. The decrease appears smoother from a H$_2$/O$_2$ ratio of 2.3.

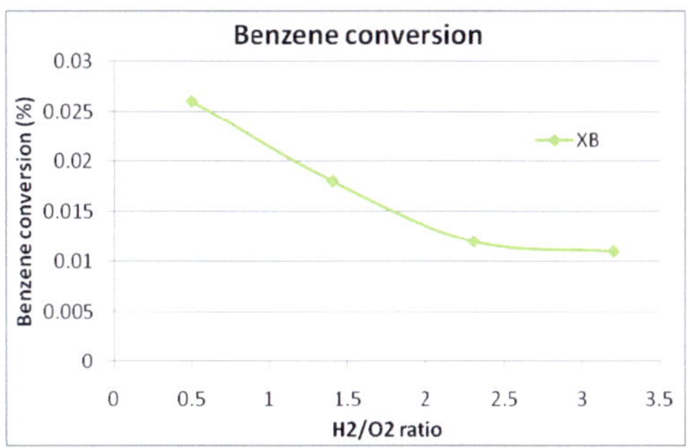

Fig. 9.1-4: Evolution of benzene conversion with respect to the H_2/O_2 ratios in reactor (exit)

9.1.2-2 Oxygen feeding mode

In order to determine whether the double-membrane dosage brings the expected benefit in terms of controlling the hydroxylation atmosphere in the reactor (effect of the oxygen membrane), an experiment was carried out in which the oxygen retentate side was shut off and the oxygen was introduced, together with benzene and nitrogen, at the reactor entrance. The used membrane foils were 50 μm $Pd_{60}Cu_{40}$ and 15 μm Ag respectively and the reaction temperature was set at 150°C for a H_2/O_2 ratio of 3.2. Here a similar phenol rate (slightly higher by 3% in co-feed mode) was obtained with a 5-fold increase in the CO_2. A similar increase, i.e. 5-fold, was also observed for the benzene conversion rate.

The phenol selectivity appeared to be 5 times lower in co-feed mode, which clearly shows the important and positive influence of oxygen dosage, via a second membrane, in the system. This is a crucial observation as it confirms the benefit of using 2 membranes for the hydroxylation performance. A significant influence on the phenol yield is, however, not noticed as the benzene conversion and phenol selectivity balance each other out.

As previously reported [26, 27], the direct introduction of oxygen at the reactor entrance reduces the available active membrane area for hydroxylation.

The data in table 9.1-1 compare the values obtained in oxygen co-feed mode at the reactor entrance with those obtained in dosage mode via the silver foil.

Table 9.1-1: Performance comparison between oxygen co-feed mode and double membrane dosage at T=150°C, P_{H2}=0.06 bar and H_2/O_2=2.3.

Oxygen supply mode	Dosage via 2nd membrane	Co-feed at reactor entrance
Phenol rate (x10^{-10} mol/min)	1.37	1.42
CO_2 rate (x10^{-10} mol/min)	91.8	488
Benzene conversion (%)	0.012	0.06
Phenol selectivity (%)	8.2	1.7

9.1.2-3 Temperature and membrane type

Other experiments, at reaction temperatures of 150°C and 180°C respectively, were carried out using plated composite membranes (5 µm $Pd_{59}Cu_{41}$ and 5 µm Ag) instead of foils in order to check the influence of the different reaction parameters. The surface of the produced PdCu composite membrane was not activated by any additional deposition of active Pd, as was the case for the PdCu foil. The Pd-based layer, which was produced as previously described by electroless plating, features a surface state with crystallites. The possibility exists that defects, which result in hydrogen diffusing both in dissociated and molecular form, are present in this layer; this has been ascertained by the experimental permeation behavior of hydrogen through the $Pd_{59}Cu_{41}$ membrane produced by ELP.

The morphology of the cold rolled surface of the PdCu foil does not seem to favor the adsorption of benzene as no product formation was detected prior to surface seeding with palladium particles (no phenol and only traces of CO_2). Water was produced, which means that a direct reaction between H_2 and O_2 did occur on the untreated PdCu foil surface. The presented measurements have been realized for the same H_2/O_2 ratio.

Fig. 9.1-5 shows the phenol and CO_2 production rates in relation to the temperature and used membrane type (50 µm $Pd_{60}Cu_{40}$ / 15 µm Ag foils and 5 µm $Pd_{59}Cu_{41}$ / 5 µm Ag composite membranes). Note that the phenol rates were multiplied by a factor of 100 so that they could be presented together with CO_2 on the same scale.

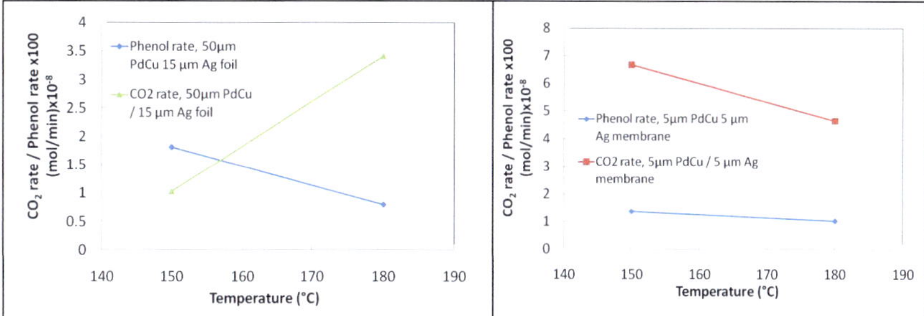

Fig. 9.1-5: Phenol and CO_2 production rates with respect to the reaction temperature and membrane type, for a same H_2/O_2 concentration ratio in the reactor.

Three points of interest have to be mentioned here:

- Higher product rates are obtained overall where composite membranes, and not the foils, are used in the reactor. This may be due to a higher catalytic activity of the composite membrane surface, which allows it to hydroxylate benzene into phenol and also oxidize it to produce CO_2. Whereas the alloy compositions are similar, the result is a higher surface roughness in the case of the composite membrane; this could lead to a higher surface area, which benefits the involved reactions.

- The phenol rate decreases with increasing temperature. This may be due to an accelerated phenol oxidation at higher temperatures.

- The CO_2 rates surprisingly exhibit different behaviors at 180°C when the composite membranes are used. A higher CO_2 rate is expected at 180°C than at 150°C (the temperature favors the oxidation of C-species), as is observed over the PdCu foil. This behavior could be related to the use of the 5 μm Ag plated membrane for oxygen dosage, which may result in an irregular delivery of O_2 at higher temperature due to the presence of pores and defects in the plated layer.

9.1.2-4 Summary of experimental results for parameter influence determination

Table 9.1-2 summarizes the obtained experimental results in order to be able to compare the influence of the process parameters on the benzene hydroxylation reaction. The influence of the H_2/O_2 concentration ratios in the reactor is, in particular, focused on here, as this is presented as one of the most important parameters in the published works [26, 27]. An overview on the influence of the reaction temperature and membrane type (foils and plated composite membranes) is also found here. Finally, an experiment realized by co-feeding oxygen, together with benzene, at the reactor entrance confirms the benefit of dosing oxygen via a second membrane in order to improve the phenol selectivity.

Overall, the highest phenol selectivity was measured, using the foils, at 150°C with a value approaching 10% [61]. This initial result was encouraging, as a phenol selectivity value under 5% was reported by DSM/ECN for the same reaction [29]. The selectivity values obtained for other temperatures and membranes were stable at around 1.3%.

The calculated benzene conversion resulted in very low values, under 0.1%; this is related to the high stability of the benzene in the gas phase, which was observed during all the realized experiments.

Table 9.1-2: Parameters used and reaction performance measured for the first experiment series aiming to establish the influence of several process parameters on benzene direct hydroxylation. The benzene partial pressure for these experiments was set at 0.14 bar (temperature of 0°C in cooling trap).

Series of exp.	Membrane type	O_2 feed type	T (°C)	P_{H2} (bar)	H_2/O_2	$X_{Benzene}$ (%)	Phenol rate (mol/h)	Phenol selectivity (%)	CO_2 selectivity (%)
S1	Pd/PdCu foil	Ag-foil dosage	150	0.04	0.5	0.026	6.4×10^{-9}	3.4	96.6
S1	Pd/PdCu foil	Ag-foil dosage	150	0.05	1.4	0.018	1.3×10^{-8}	9.6	90.4
S1	Pd/PdCu foil	Ag-foil dosage	150	0.04	1.8	0.004	1.1×10^{-8}	9.5	90.5
S1	Pd/PdCu foil	Ag-foil dosage	150	0.06	2.3	0.012	8.2×10^{-9}	8.2	91.8
S1	Pd/PdCu foil	Ag-foil dosage	150	0.07	3.2	0.011	7.4×10^{-9}	7.9	92.1
S1	Pd/PdCu foil	Ag-foil dosage	180	0.11	3.2	0.04	4.7×10^{-9}	1.4	98.6
S2	PdCu / Ag composite mem.	Ag-foil dosage	150	0.13	1.8	0.08	8.2×10^{-9}	1.2	98.8
S2	PdCu / Ag composite mem.	Ag-foil dosage	180	0.13	3.2	0.06	6.0×10^{-9}	1.3	98.7
S1	Pd/PdCu foil	co-feed	150	0.06	2.3	0.06	8.5×10^{-9}	1.7	98.3

9.1.2-5 Introduction to the use of surface-modified PdCu foils for catalytic improvement purposes

Based on the initial concept from Niwa et al. [24], the direct conversion of benzene into phenol over a Pd-based membrane was reproduced using the double-membrane reactor. While the reaction did occur as suggested by the authors, i.e. using the gas phase route, the system is, however, dominated by side-reactions with a phenol C-based selectivity, which only reaches about 10%. An improvement of the phenol selectivity in the reactor is required under these conditions. Several catalytic systems were deposited onto the surface of the Cu foils with the aim of achieving a higher reactivity of the membrane to phenol. The 50 µm PdCu foils were used as substrates; this membrane type was characterized in uncoated state and featured no apparent defects (unlike the plated membranes), which could lead to an undetermined dosage of hydrogen in the reactor. Previous experiments, where Pd-seeded PdCu foils were used, have shown that a temperature of 150°C resulted in the best phenol selectivity. For this reason, a lower range of temperatures should be privileged for the series of experiments with the different catalytic systems. In addition to the catalytic activity of the used modified foil, the influence of the process parameters, e.g. the temperature and partial pressure amplitude of the reaction gases, was also studied. The results are presented in the following paragraphs and have also been published [80]. Due to the variation of several parameters from one experiment to another, a direct comparison between the systems is not always possible. One can, however, assess their influence on the reaction performance, i.e. the phenol formation rate and selectivity as well as the rate of undesired reaction products, compared to a system without surface modification.

9.2. PdAu as catalytic surface

9.2.1. Effect of the catalytic modification

In this series of experiments, with PdAu as the catalytic layer on top of the membrane, the reaction temperature was kept at 150°C and the benzene partial pressure in the feed was kept constant at 0.04 bar. The hydrogen partial pressure was 0.05 bar, while the oxygen partial pressure was varied. The best previously obtained performance under these conditions, with the Pd-seeded PdCu foil, was at a H_2/O_2 ratio of 1.4. The phenol rate was 1.3×10^{-8} mol/h, while the associated selectivity was 9.6%, i.e. the remaining selectivity of the C-products, in this case the CO_2, was 90.4%. Table 9.2-1 shows the experimental results obtained under these conditions for the PdAu layer. The UF layer was used, instead of the Ag foil, for oxygen dosage in the reaction area, as potential defects in the foil could, otherwise, lead to an undefined distribution of oxygen in the reactor.

Table 9.2-1: Parameters and results of the experiments performed at constant temperature with the 1 µm PdAu-modified foil.

Series of exp.	Surface modifica-tion type	O_2 feed type	T (°C)	P_{H2} (bar)	H_2/O_2	$X_{Benzene}$ (%)	Phenol rate $x10^{-7}$ (mol/h)	Phenol selectivity (%)	CO_2 rate $x10^{-6}$ (mol/h)	H_2O Rate $x10^{-3}$ (mol/h)
S1	Pd/PdCu foil	Ag-foil dosage	150	0.05	1.4	0.018	0.13	9.6	0.73	/
S3	1 µm PdAu	UF-TR.	150	0.05	1.4	0.022	5.4	58.4	2.3	7.0
S3	1 µm PdAu	UF-TR.	150	0.05	1.0	0.023	7.2	67.6	2.1	8.5
S3	1 µm PdAu	UF-TR.	150	0.05	0.8	0.023	7.0	63.6	2.4	10.0
S3	1 µm PdAu	UF-TR.	150	0.05	0.7	0.023	6.2	57.8	2.7	6.5

One can immediately recognize that side-reactions are still present and have a significant influence in the system, in particular the water formation, whose rate is 3 to 4 orders of magnitude higher ($1x10^{-3}$ mol/h range) than the CO_2 or phenol rates. The large predominance of water illustrates the much quicker kinetics of direct hydrogen oxidation to H_2O over Pd-based surfaces vis-à-vis any other reaction. The utilization of active hydrogen for benzene hydroxylation to phenol, in the reactor, appears to be very poor. In comparison to the initial experiments without catalyst modification, the CO_2 rate is approx. twice as high for the PdAu layer, in the range of $2x10^{-6}$ mol/h, while also featuring a higher activity in terms of undesired oxidation reactions.

It can also be clearly seen, however, that the PdAu catalytic surface led to an improvement in the phenol rate and selectivity, the latter being much higher, at above 50%, than the previously obtained 9.6%; this makes phenol the main C-product from benzene which exits the reactor. Selectivities ranging from 57.8% to 67.6% were measured for H_2/O_2 ratios, which varied between 0.7 and 1.4. The stable parameters used in the experiments, performed over PdAu, also make it possible to visualize the effect of the H_2/O_2 concentration ratio, in the reactor, on the phenol rate and selectivity; this is plotted in Fig. 9.2-1. A maximum in terms of the phenol rate and selectivity for a H_2/O_2 ratio close to 1 was also observed here. A decrease in both the rate and selectivity to phenol can be detected as the atmosphere becomes more oxidative or hydrogenative.

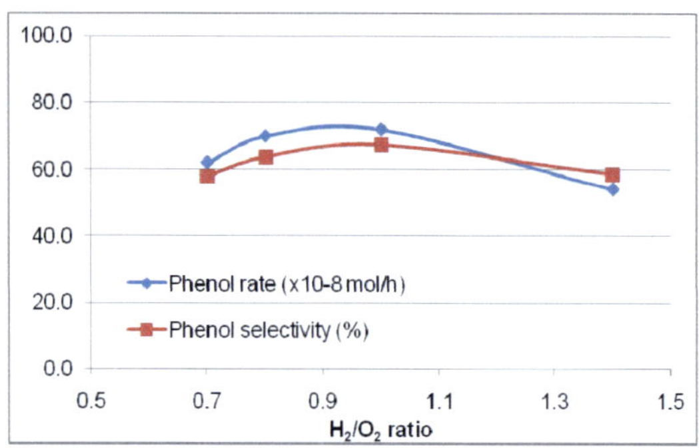

Fig. 9.2-1: Evolution of phenol rate and selectivity with the H_2/O_2 ratio in the reactor for the PdAu-modified foil.

In comparison to the experiments without surface modification, the phenol rate has dramatically improved by a factor of 55, which shows the large extent to which the PdAu surface favors the phenol formation kinetics. This could be related to the enhanced formation of active surface species, presumably OH-radicals, from gas phase oxygen and dissociated hydrogen species on the catalyst surface, which, in turn, convert readily adsorbed benzene molecules into phenol. It was not possible to quantify the water rate during the experiments with PdCu, the values now obtained, however, show that water is by far the main product of the system, with formation rates between 6.5 $\times 10^{-3}$ mol/h and 1×10^{-2} mol/h, which is more than 3 orders of magnitude that of CO_2 and 4 orders of magnitude that of phenol. These proportions suggest that water is formed on the catalyst from the direct reaction between H_2 and O_2; this reaction is much quicker than any other reaction involving benzene. This inefficient use of the gas dosed into the system for the reaction is accompanied by extremely low benzene conversions (below 0.03%); this is due to the high stability of the benzene molecules in the gas phase, in particular, at mild temperatures (low thermal activation).

9.2.2. Influence of the temperature

The influence of the temperature on the evolution of the phenol rate and reaction selectivity was investigated on a PdAu surface. It was expected that increased temperatures would activate the reaction processes and lead to higher product rates and a higher benzene conversion. The influence of the reactor temperature was studied for the $Pd_{90}Au_{10}$ layer in double-membrane mode. The investigated temperature range was between 150°C and 240°C; all other parameters were kept at constant values. In this connection, the H_2/O_2 ratio was 1. The hydrogen partial pressure at the reactor outlet was 0.05 bar at 1 bar of total pressure, while the total gas flow through the reaction channels was 110 mL/min. The benzene partial pressure in the feed was 0.04 bar.

The experimental results are presented below in Table 9.2-2 and are subsequently graphically interpreted.

Table 9.2-2: Parameters and results of the experiments performed at variable temperatures with the 1 µm PdAu-modified foil.

Series of exp.	Surface modifica -tion type	O_2 feed type	T (°C)	P_{H2} (bar)	$H_2/$ O_2	$X_{Benzene}$ (%)	Phenol rate $\times 10^{-7}$ (mol/h)	Phenol selectivity (%)	CO_2 rate $\times 10^{-6}$ (mol/h)	H_2O Rate $\times 10^{-3}$ (mol/h)
S4	1 µm PdAu	UF-TR.	150	0.08	1.0	0.023	7.0	63.4	2.43	10.1
S4	1 µm PdAu	UF-TR.	175	0.08	1.0	0.029	7.4	51.7	4.15	11.0
S4	1 µm PdAu	UF-TR.	200	0.08	1.0	0.045	8.7	38.2	8.45	11.8
S4	1 µm PdAu	UF-TR.	220	0.08	1.0	0.063	3.8	12.6	15.90	13.0

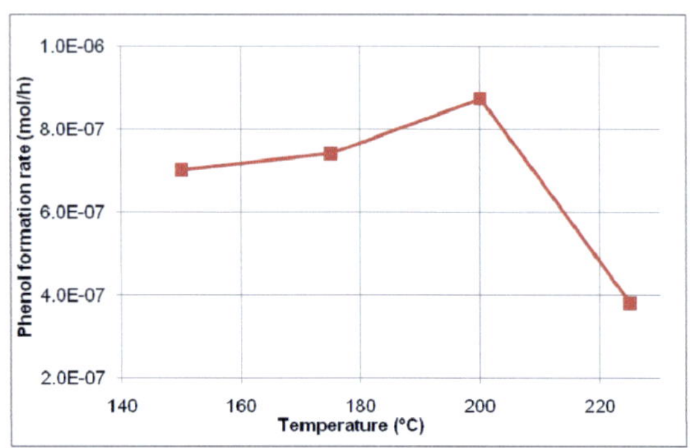

Fig. 9.2-2. Phenol formation rate as a function of the reactor temperature. Compact $Pd_{90}Au_{10}$ layer indouble-membrane operation mode (P_B: 0.04 bar; P_{H2}: 0.05 bar; P_{O2}: 0.05 bar).

Fig. 9.2-2 shows the phenol formation rate obtained as a function of the temperature. One can initially recognize that the obtained phenol rates, with the $Pd_{90}Au_{10}$ layer, are higher by more than one order of magnitude than those previously achieved with the Pd-modified PdCu foil. The phenol rates were, in fact, increased by a factor of 50 for the used set of parameters. As the temperature increases, the phenol formation rate initially, slowly increases up to a temperature of 200°C, and then drops quickly. The phenol C-selectivity, defined below in Eq. 9.2-5, decreases from 58% at 150°C to 11% at 225°C due to a consistent and more pronounced increase of the CO_2 formation rate with temperature, which is shown in Fig. 9.2-3. Figure 9.2-3 also gives the H_2O formation rate as being 3 orders of magnitude higher than the formation rate of CO_2.

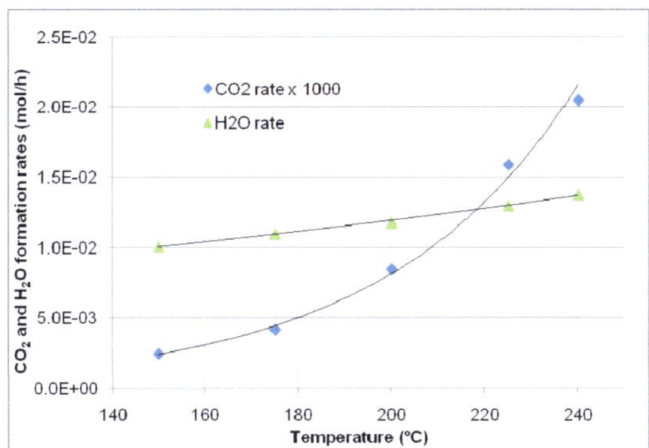

Fig. 9.2-3. CO$_2$ and H$_2$O formation rates as a function of the reactor temperature (conditions as for Fig. 9.2-2).

A higher apparent activation energy for CO_2 formation than for H_2O formation is revealed when the temperature dependency of the two formation rates in Fig. 9.2-3 are compared. If details of the true reaction mechanism, involving various surface reactions and adsorbed species [28, 29] are ignored, then the reaction network can overall be described via the following simplified parallel-consecutive reaction equations, which serves as a basis for the discussion of the kinetic performance:

$$C_6H_6 + H_2 + O_2 \rightarrow C_6H_5OH + H_2O \qquad \text{Eq. (9.2-1)}$$
$$C_6H_6 + 7.5\ O_2 \rightarrow 6\ CO_2 + 3\ H_2O \qquad \text{Eq. (9.2-2)}$$
$$C_6H_5OH + 7\ O_2 \rightarrow 6\ CO_2 + 3\ H_2O \qquad \text{Eq. (9.2-3)}$$
$$H_2 + 0.5\ O_2 \rightarrow H_2O \qquad \text{Eq. (9.2-4)}$$

In accordance with this reaction scheme, the maximum of the phenol formation rate as a function of temperature could be explained by a higher activation energy of the subsequent oxidation of phenol (9.2-3) compared to its formation (9.2-1). CO_2 formation is stoichiometrically linked to the formation of H_2O according to equations 9.2-2 and 9.2-3, but the fact that the H_2O formation rate is much higher than the CO_2 formation rate indicates that most of the H_2O

originates from the oxidation of H_2. The hydrogen-based selectivity, defined by equation 9.2-6, is thus extremely low in comparison to the carbon-based selectivity.

Carbon-based phenol selectivity:

$$S_C = \frac{R_{Ph}}{R_{Ph} + \frac{1}{6} R_{CO2}}$$

Eq. (9.2-5)

Hydrogen-based phenol selectivity:

$$S_H = \frac{R_{Ph}}{R_{Ph} + R_{H2O} - \frac{1}{2} R_{CO2}}$$

Eq. (9.2-6)

9.2.3. Influence of the partial pressures of hydrogen and oxygen

The effects of the partial pressures of hydrogen and oxygen on the reactions on the $Pd_{90}Au_{10}$ layer were investigated in co-feed mode and not where these two reactants are supplied via membranes in order to overcome restrictions in the variation of the partial pressures, which arise in the case of gas dosage via membranes. The conditions were chosen so that the hydrogen and oxygen partial pressures did not change by more than 10% along the reactor. It must, however, be pointed out that the conditions on the surface of the catalytic layer in hydrogen co-feed mode might be different than those observed in the case of hydrogen supply through the catalytic layer. This is because the rate of hydrogen adsorption from the gas phase is, in general, not the same as the rate of hydrogen diffusion through the membrane. This therefore may affect the interplay of the involved surface processes (adsorption and desorption of various reactants and products as well as the surface reactions of different intermediates).

The reaction temperature during these experiments was kept constant at 150°C. The partial pressures were varied between 0.05 and 0.18 bar in case of oxygen and between 0.05 and 0.37 bar in case of hydrogen. The benzene partial pressure in the reactor was set at 0.04 bar. Total pressure and flow through the reaction channels were similar to the conditions of the previous experiment series. The experimental data collected in this series can be found below in Table 9.2-3.

Table 9.2-3: Parameters and results of the experiment series performed with varied partial pressures using PdAu modified membranes.

Series of exp.	Surface modifica-tion type	O_2 feed type	T (°C)	P_{H2} (bar)	H_2/O_2	$X_{Benzene}$ (%)	Phenol rate $x10^{-8}$ (mol/h)	Phenol selectivity (%)	CO_2 rate $x10^{-6}$ (mol/h)	H_2O Rate $x10^{-3}$ (mol/h)
S5	1 µm PdAu	co-feed	150	0.05	0.10	0.022	3.1	6.3	2.8	6.00
S5	1 µm PdAu	co-feed	150	0.16	0.10	0.022	4.3	8.7	2.7	6.50
S5	1 µm PdAu	co-feed	150	0.26	0.10	0.022	2.4	5.1	2.7	7.00
S5	1 µm PdAu	co-feed	150	0.37	0.10	0.022	1.1	2.3	2.8	7.50
S6	1 µm PdAu	co-feed	150	0.25	0.05	0.022	0.6	1.6	2.3	6.85
S6	1 µm PdAu	co-feed	150	0.25	0.10	0.022	2.4	5.1	2.7	6.95
S6	1 µm PdAu	co-feed	150	0.25	0.15	0.022	1.9	3.5	3.1	7.00
S6	1 µm PdAu	co-feed	150	0.25	0.18	0.022	1.7	3.0	3.3	7.05

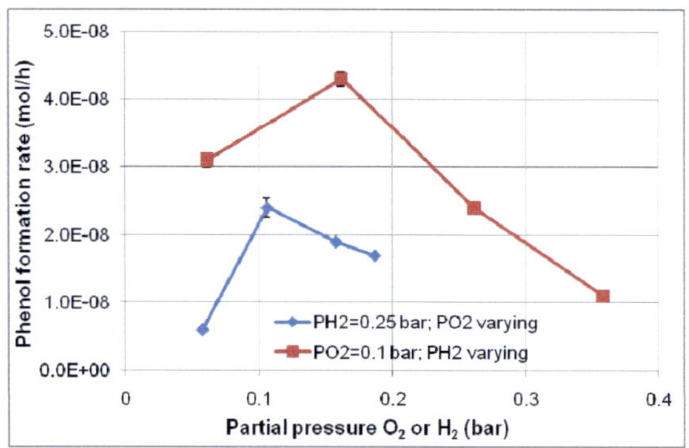

Fig. 9.2-4. Phenol formation rate versus P_{H2} and P_{O2}. Compact $Pd_{90}Au_{10}$ layer in co-feed operation mode (P_B: 0.04 bar; T: 150°C).

Figure 9.2-4 shows the phenol formation rate measured for varying oxygen partial pressures at constant P_{H2} and varying hydrogen partial pressures at constant P_{O2}.

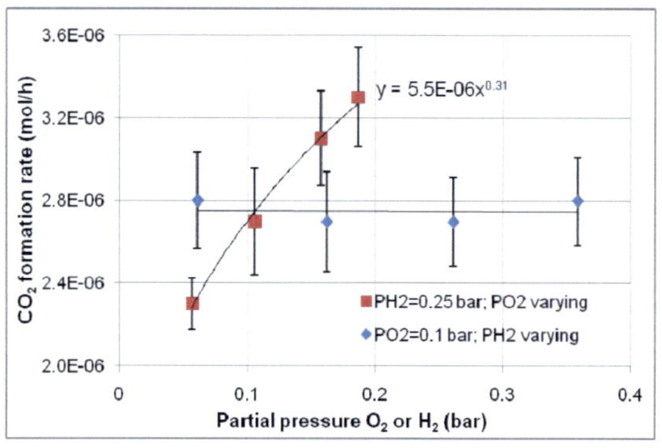

Fig. 9.2-5. CO_2 formation rate versus P_{H2} and P_{O2} (conditions as for Fig. 9.2-4).

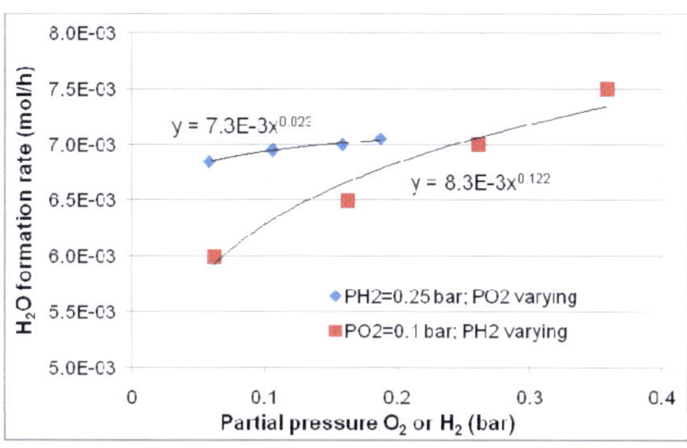

Fig. 9.2-6. H_2O formation rate versus P_{H2} and P_{O2} (conditions as for Fig. 9.2-4).

Figure 9.2-5 shows the rate of CO_2 formation, while Fig. 9.2-6 illustrates the H_2O formation rate. Similar to the variation of temperature shown in Fig. 9.2-2, a maximum is also observed for the variation of the H_2/O_2 ratio. The optimum value lies between 1.2 and 1.4. The occurrence of a maximum in the phenol formation rate at a H_2/O_2 ratio of 1.4 was already observed in our previous study [13]. Figure 9.2-5 shows that the CO_2 formation rate is virtually unaffected by the partial pressure of hydrogen, whereas it increases at a power of 0.31 with the oxygen partial pressure. Fig. 9.2-6 shows that the formation rate of H_2O is only weakly influenced by the oxygen ($nO_2 = 0.023$) and hydrogen ($nH_2 = 0.12$) partial pressures within the covered range of partial pressures.

One important question, concerning the formation of CO_2, in the parallel-consecutive reaction scheme, presented in 9.2-1 to 9.2-4 equations, is whether CO_2 is mainly formed from benzene, i.e. in a parallel reaction, or from phenol, i.e. in a consecutive reaction. The latter would basically explain the observed maximum of the phenol rate versus the oxygen partial pressure at constant P_{H2} (where the consecutive oxidation step has a higher partial reaction order for oxygen). This question is addressed in Fig. 9.2-7 where the formation rate of CO_2, scaled with the benzene partial pressure for the experiments with varying P_{O2} (already shown in Fig. 9.2-5), is compared with the formation rate of CO_2 from 3 additional experiments involving the co-feeding of phenol and oxygen in the absence of benzene, similarly scaled with the phenol partial pressure. The latter experiments were all realized at a temperature of 150°C with a phenol partial pressure of 1.38 mbar and an oxygen partial pressure varying between 0.12 bar and 0.18 bar. The total pressure was set at 1 bar.

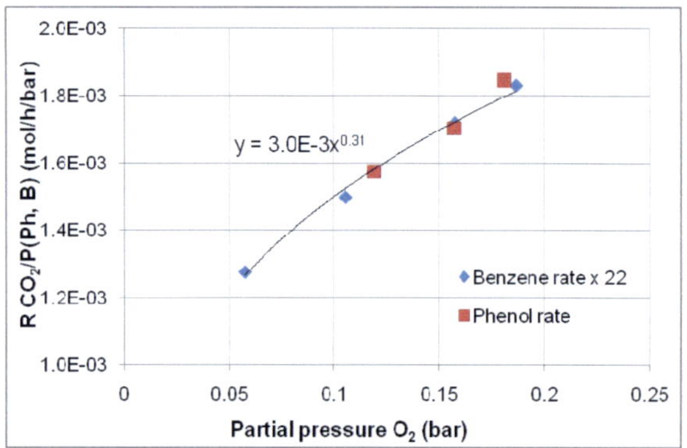

Fig. 9.2-7. CO_2 formation rate scaled with the partial pressure of the aromatic feed component (P_B or P_{Ph}) as a function of P_{O2}. Comparison of benzene and phenol oxidation on a $Pd_{90}Au_{10}$ layer in co-feed operation (P_B: 0.04 bar or P_{Ph}: 1.38 mbar; P_{H2} = 0.1 bar; T: 150°C).

By scaling the CO_2 formation rate, using the partial pressure of the aromatic feed component, it is possible to correct the influence of the different partial pressures used in the two series of experiments. The oxygen dependency of the CO_2 formation can be isolated by assuming the same dependency of the CO_2 formation rate, on the partial pressure of the aromatic feed component, for benzene and phenol, e.g. first order. Interestingly, almost the same oxygen dependency is observed for benzene and phenol oxidation, the rate is, however, 22 times higher for phenol. The fact that the oxidation of phenol was performed here in the absence of hydrogen, while hydrogen benzene was present in the experiments, does not interfere with this interpretation, as can be seen from Fig. 9.2-5, which shows that the influence of the hydrogen partial pressure on the rate of CO_2 formation is minimal.

The fractional reaction orders for oxygen and hydrogen as well as the maximum of the phenol rate versus the H_2/O_2 ratio clearly signals that the reaction is kinetically much more complicated than suggested by the simple reaction scheme of equations 9.2-1 to 9.2-4.

9.3. V_2O_5/PdAu as catalytic surface

9.3.1. Effect of the catalytic modification and comparison with PdAu

This series of experiments, involving the V_2O_5/PdAu catalytic system, were carried out under the same conditions as for PdAu, i.e. a reaction temperature of 150°C and a benzene partial pressure in the feed of 0.04 bar. The aim, in this connection, was to increase the surface area for the reaction. The hydrogen partial pressure was kept constant at 0.04 bar and the oxygen partial pressure was varied for two experiments. It was then increased to 0.08 bar and the oxygen partial pressure was varied for another two experiments. Table 9.3-1 shows the experimental results obtained under these conditions with the PdAu layer.

Table 9.3-1: Parameters and results of the experiments performed at constant temperature with the V_2O_5/PdAu-sputtered foil.

Series of exp.	Surface modification type	O_2 feed type	T (°C)	P_{H2} (bar)	H_2/O_2	$X_{Benzene}$ (%)	Phenol rate $x10^{-7}$ (mol/h)	Phenol selectivity (%)	CO_2 rate $x10^{-6}$ (mol/h)	H_2O Rate $x10^{-3}$ (mol/h)
S7	V_2O_5/PdAu	UF-TR.	150	0.04	0.6	0.022	2.8	31.2	3.7	7.5
S7	V_2O_5/PdAu	UF-TR.	150	0.04	1.2	0.022	4.8	53.3	2.5	9.0
S7	V_2O_5/PdAu	UF-TR.	150	0.08	1.2	0.022	4.0	63.1	1.4	6.6
S7	V_2O_5/PdAu	UF-TR.	150	0.08	1.4	0.022	5.4	59.8	2.2	7.3

As for PdAu, one observes an increase in both the phenol rate and selectivity in comparison to the PdCu foil before modification. The phenol rate has improved by a factor of 20 to 40 for a H_2/O_2 atmospheric ratio between 0.6 and 1.4; 5.4×10^{-7} mol/h was obtained for a ratio of 1.4. The selectivity varies between 31.2% (CO_2 still dominating) and 63.1% (phenol being the main C-product from the reactor) for the same set of parameters. The peak selectivity was reached for a H_2/O_2 ratio of 1.2. This observation is in accordance with what was obtained over PdCu and PdAu, with optimum selectivity at 1.4 and 1.0 respectively. The improvement in the reaction performance is, unexpectedly, not as high as with the PdAu layer-based catalyst, where phenol was always dominant. The phenol rate over V_2O_5/PdAu is approx. 20% less important than over PdAu. The increase in the active catalytic surface, via the deposition of PdAu particles over the insular V_2O_5 support, thus did not result in the desired surface state (see the characterization for the V_2O_5/PdAu catalyst). A thin PdAu layer was deposited in a short sputtering period, instead of the particular morphology; this reduced the surface increase in comparison to the layer deposition on the flat substrate. The oxide particles, distributed on the PdCu foil, may possibly influence the surface dispersion of the active hydrogen species emerging from the bulk of the PdCu foil; this could potentially lead to a less homogeneous distribution on the PdAu catalyst. This could be an explanation for the less optimized performance in comparison to PdAu, even though the active surface has been increased by a certain extent. In the case of the formation rate for the byproducts, both the CO_2 and water are produced in similar proportions to that observed for PdAu. The CO_2 rate varied in a wider range between 1.4×10^{-6} mol/h and 3.7×10^{-6} mol/h (in comparison to a range of 2.1×10^{-6} mol/h to 2.7×10^{-6} mol/h over PdAu), while the water rate changed from 6.6×10^{-3} mol/h to 9.0×10^{-3} mol/h (6.5×10^{-3} mol/h to 10.0×10^{-3} mol/h over PdAu). No significant change was observed between the two catalytic systems at 150°C, even though the process parameters were not strictly identical in the two series of experiments (P_{H2} ranged between 0.04 bar and 0.08 bar for the S7 series of experiments, whereas it was fixed at 0.05 bar for S3).

9.3.2. Influence of the temperature and comparison with PdAu

More experiments were carried out at increasing reaction temperatures, between 150°C and 225°C, in order to compare the activity of the biphasic metal/ceramic catalyst with that of the system solely based on PdAu. All experiments were carried out under constant atmosphere in the reactor (benzene partial pressure of 0.04 bar, H_2 partial pressure of 0.08 bar with a H_2/O_2 ratio of 1.2). The results are summarized in Table 9.3-2.

Table 9.3-2: Experimental data collected varying the temperature over the V_2O_5/PdAu catalytic system.

Series of exp.	Surface modifica-tion type	O_2 feed type	T (°C)	P_{H2} (bar)	H_2/O_2	$X_{Benzene}$ (%)	Phenol rate $x10^{-7}$ (mol/h)	Phenol selectivity (%)	CO_2 rate $x10^{-6}$ (mol/h)	H_2O Rate $x10^{-3}$ (mol/h)
S8	V_2O_5/Pd Au	UF-TR.	150	0.08	1.2	0.022	5.4	59.8	2.16	7.30
S8	V_2O_5/Pd Au	UF-TR.	175	0.08	1.2	0.030	5.8	47.5	3.85	8.10
S8	V_2O_5/Pd Au	UF-TR.	200	0.08	1.2	0.045	7.0	34.2	8.14	9.00
S8	V_2O_5/Pd Au	UF-TR.	225	0.08	1.2	0.064	5.2	13.5	20.0	10.5

The phenol formation rates as a function of the temperature are shown in Fig. 9.3-1. For comparison purposes, the rates observed with PdAu also appear in Fig. 9.3-1.

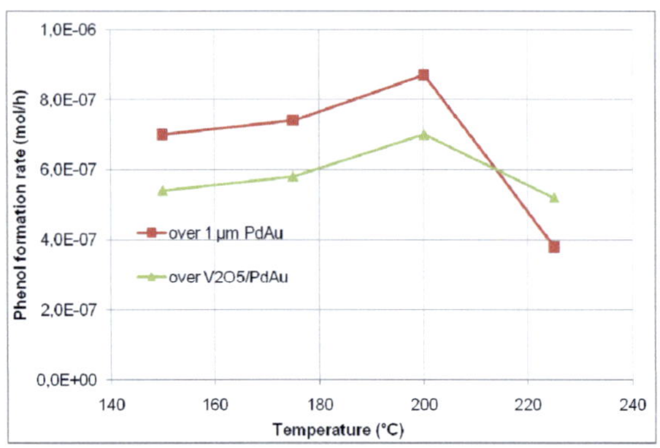

Fig. 9.3-1. Phenol formation rate as a function of the reactor temperature.

A similar profile over V_2O_5/PdAu is observed as for the PdAu layer. The phenol rate initially increases with the temperature and then decreases after 200°C due to the increase in the rate of phenol oxidation to CO_2. As observed at constant temperature, the phenol rate for the V_2O_5/PdAu catalytic surface is about 20% less important than with PdAu in the ascending domain (conditions favoring benzene hydroxylation to phenol in comparison to its oxidation to CO_2). The low surface increase, when compared to the original idea, due to the insular microstructure based on V_2O_5 and the influence of the latter on active species distribution on the surface is proposed as a possible explanation for the lower phenol rate. At higher temperatures, the decrease in the phenol rate, due to the parallel oxidation process, appears less severe with V_2O_5/PdAu. The higher temperature domain was not privileged for these experiments due to the better selectivity observed in the low temperature range, which is due to a limited activation of the oxidation reaction of the C-species.

Fig. 9.3-2 compares the by-product formation rates (CO_2 and H_2O) with each other over the 2 PdAu-based catalysts.

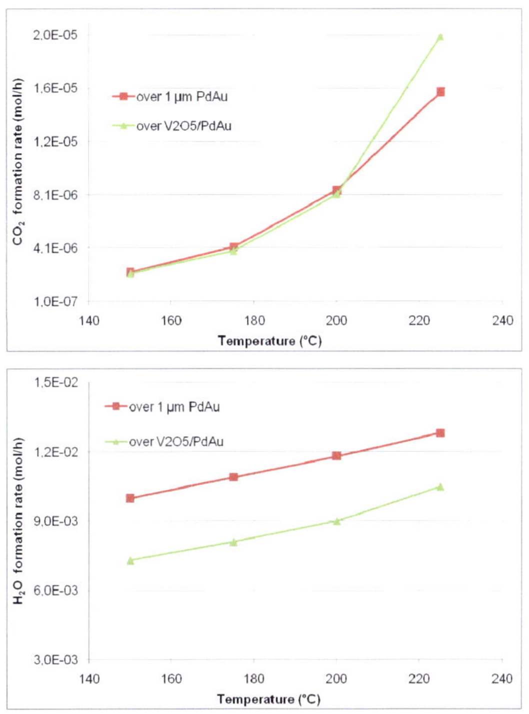

Fig. 9.3-2. CO_2 and H_2O formation rates as a function of the reactor temperature.

With regard to the CO_2 formation, both rates have exactly the same profile with a slightly quicker rate over V_2O_5/PdAu in the high temperature range. This was expected as the reactions, which do not involve hydrogen diffusing through the different catalyst types should not be influenced by them, seeing as the surface itself is fully PdAu-based and the increase in surface area was not significant enough. Both oxygen and benzene adsorb on the PdAu catalyst from the reaction side. In this case, the water rate seems to be influenced by the catalyst type. As for the phenol rate, the PdAu layer shows an approx. 20% higher rate than with V_2O_5/PdAu for all investigated temperatures. The presence of the V_2O_5 particles on the foil, which could result in less homogeneity in the distribution of active hydrogen species on the PdAu surface, could again be the reason for this phenomenon. In the investigated temperature range, a nearly 10 fold increase in the CO_2 rate over the catalytic surface is observed, while the water has not doubled in the same interval. This indicates a much higher activation energy for the oxidation process of the C-products in comparison with that of the (much faster) water reaction from H_2 and O_2.

9.4. PdGa as catalytic surface

The intermetallic phase $Pd_{50}Ga_{50}$ at.% is the third catalytic system tested in this project. The specificity of its surface state, with Pd atoms isolated from each other by inert Ga atoms, is expected to play a role in the adsorption behavior of benzene and thus influence the rate of the reactions involving benzene. PdGa layers with several thicknesses (5 µm, 1.4 µm and 0.4 µm) were produced on PdCu foils, as the 5 µm PdGa-based membrane showed a much lower hydrogen permeability than the other surface-modified membranes (see characterization chapter of the catalytic systems). The thickness of the catalytic film had to be proportionally reduced in order to obtain the same atmospheric conditions in the reaction channels (similar hydrogen partial pressure for all systems). 3 PdGa-modified foils were used in the reactor. In addition to the comparison with the other 2 catalyst types, a comparison between their respective performances in relation to the thickness of the used catalytic film will also be presented at the end of this section.

9.4.1. 5.0 µm PdGa membrane

9.4.1-1. Effect of the catalytic modification and comparison with the previous systems

The 5.0 µm PdGa membrane was initially tested in the reactor. The lower hydrogen permeability observed during the characterization of this membrane in comparison to the other surface-modified foils, even at the maximum driving force, which can be achieved by the setup (3 bar of hydrogen retentate pressure), resulted in very low hydrogen partial pressures in the reaction area, this is because all gases were dosed through the PdGa-modified membrane. 7 mbar of hydrogen was only reached at 150°C, which is more than 5 times less than what is dosed through PdAu for the same set of process parameters. Experiments in co-feed mode, where hydrogen is introduced together with benzene, nitrogen and oxygen at the reactor entrance, thus had to be performed in order to obtain the same range of gas partial pressures at the reactor exit as those possible with the more permeable membranes. The results of experiments carried out in membrane dosage mode and co-feed are presented in Table 9.4-1.

Table 9.4-1: Experimental data collected at constant temperature over the 5 μm PdGa catalytic system.

Series of exp.	Surface modifica -tion type	O_2 feed type	T (°C)	P_{H2} (bar)	$H_2/$ O_2	$X_{Benzene}$ (%)	Phenol rate $x10^{-8}$ (mol/h)	Phenol selectivity (%)	CO_2 rate $x10^{-7}$ (mol/h)	H_2O Rate $x10^{-3}$ (mol/h)
S9	5 μm PdGa	Ag-foil dosage	150	0.007	0.5	0.002	0.36	24.9	6.5	1.1
S9	5 μm PdGa	Ag-foil dosage	150	0.007	1.4	0.002	1.6	26.0	2.7	0.9
S9	5 μm PdGa	co-feed	150	0.14	1.4	0.005	11	54.5	5.5	3.8
S9	5 μm PdGa	co-feed	150	0.20	2.0	0.003	3.0	50.0	1.8	4.5

An improvement in the phenol selectivity, in comparison to the PdCu foil, is observed with values around 25% (improvement by a factor 2.5) at low hydrogen partial pressures (0.007 bar) and for a H_2/O_2 ratio ranging from 0.5 to 1.4. Even though the phenol selectivity has improved, the phenol rate is comparable to that before modification, with values ranging from 0.36 $x10^{-8}$ mol/h to $1.6x10^{-8}$ mol/h (whereas $1.3x10^{-8}$ mol/h was obtained over PdCu). This indicates that the catalyst was more effective at slowing down the CO_2 formation than at improving the phenol kinetic. The average CO_2 rate measured in this case was, in fact, $4.6x10^{-7}$ mol/h, whereas the previous minimal rate measured among the different membranes tested (PdCu, PdAu and V_2O_5/PdAu) has been 7.3×10^{-7} mol/h. The active-site-isolation effect of the catalyst surface, which was expected for the benzene adsorption behavior, may play a role in this connection. A weaker bond between the benzene molecule and an isolated Pd atom, in comparison to the stronger adsorption on 3 neighboring Pd atoms, should lead to a quick desorption and thus give the O species less time to attack the benzene and convert it to CO_2. In accordance with the observed stable phenol rates, the obtained presumably weaker bond does not influence the occurring benzene hydroxylation. The water formation rates are also lower than those obtained with the other catalysts; an average value of 1.0×10^{-3} mol/h was measured here. This is 6 times less than that obtained over PdAu. One must, however, bear in mind that both the hydrogen and oxygen partial pressures in these experiments are much lower than with the other catalytic systems due the restricted hydrogen permeability of the modified membrane. Experiments in co-feed performed at larger partial pressure amplitudes were thus required in order to be able to determine the catalyst effect.

Two experiments were carried out at $P_{O2}=0.1$ bar and $P_{H2}=0.14$ bar and 0.20 bar respectively. Higher hydrogen partial pressures, measured at the reactor end, were applied in co-feed. Under these conditions, the product rates were increased on average by approx. a factor of 8 for phenol and by a factor of 4 for water. The CO_2 rate remained stable at about 3.8×10^{-7} mol/h. The absence of an increase in the CO_2 rate, in spite of a much higher oxygen partial pressure in the reactor (14 times increase), correlates the hypothesis that the presence of more oxygen species around the absorbed benzene molecule is not decisive for the reaction due to the assumed shorter residence of benzene on the catalyst surface. This adsorption behavior is more beneficial to the hydroxylation kinetic than that of the oxidation. Phenol was the dominant C-product, with selectivities of 50% and above, in the case of the experiments carried out in co-feed mode. While still by far the dominant product in the system, the water kinetic appears to have been slowed down to some extent by the PdGa surface, as the measured rates (approx. 4.2×10^{-3} mol/h) are less than twice that obtained over PdAu under less oxidative conditions (P_{O2} twice to 3 times lower). The presence of spacing Ga atoms between the catalytically active Pd atoms also appears to influence the oxidation of hydrogen in a positive manner for the reaction under observation here. The observations realized in membrane dosage and co-feed modes are similar in this connection. While the $Pd_{50}Ga_{50}$ at.% surface does not significantly improve the phenol kinetics, it does play a key role in slowing down the side-reactions, which is, of course, of interest in such a system as this where a dramatic reduction in the influence of the side-reactions is desired.

The differences in the reaction performance can be observed when the successive experiments, carried out via membrane dosage and co-feed over PdAu and PdGa respectively, are compared. Where the behavior is the same over PdGa, the PdAu layer exhibits different behaviors - in spite of the different partial pressure amplitudes - regardless of whether the hydrogen is transported through the membrane or in the gas phase. The highest phenol rates and selectivities are observed in double-membrane dosage over PdAu, while co-feed experiments over the same surface surprisingly result in limited phenol rates, in the same range as for PdGa, and much lower selectivities, i.e. below 10% (see Table 9.2-3). PdAu, as the catalyst, dissociates hydrogen and supplies active surface species for the benzene hydroxylation regardless of the hydrogen supply mode. The amount of active hydrogen species adsorbed on the PdAu surface and able to react with benzene in a co-feed dosage of the gas, however, appears in this case to be much lower than when homogeneously dosed through the membrane. The high amount of CO_2 and water produced in co-feed mode over the PdAu surface shows that all species are present in proportions similar to double-membrane dosage. Surface hydrogen species appear to be used less efficiently when coming from the gas phase, as they react more with oxygen to form water than with the adsorbed benzene. This was not observed with the PdGa catalyst.

9.4.1-2. Verification of the balance of the system in a co-feed experiment

The balance of the system can be performed in order to ensure that all the species involved in the system and associated formation reactions have been considered and detected. The amount of gas dosed through the membranes during the hydroxylation experiments cannot be precisely assessed and differs to some extent from that obtained during the permeation measurements due to the combined dosage of the several species, which changes the partial pressures conditions. The balance was thus performed using experiments realized in co-feed mode. An example, with the 5 µm PdGa modified foil at 150°C and at a H_2/O_2 ratio of 1.4, is provided below.

The balance on the C, H, O and N elements involves the examination of the phenol formation from benzene, water formation from H_2 and O_2 and the CO_2 formation from the C_6 species. The denomination F represents the entrance flow (in mol/h) of the dosed gas or the production rate or exit flow of the species detected out of the system (also in mol/h) respectively.

C-balance:

C in (mol/h) = 6 x F Benzene gas in (mol/h)

C out (mol/h) = 6 x F Benzene solid out (mol/h) + 6 x F Phenol solid out (mol/h) + F CO_2 gas out (mol/h)

The balance imposes C in = C out as there is no accumulation in the reactor.

H-balance:

H in (mol/h) = 2 x F H_2 in (mol/h) + 6 x F Benzene gas in (mol/h)

H out (mol/h) = 2 x F H_2 out (mol/h) + 6 x F Benzene solid out (mol/h) + 6 x F Phenol solid out (mol/h) + 2 x F H_2O solid out (mol/h)

Balance condition: H in = H out

O-balance:

O in (mol/h) = 2 x F O_2 in (mol/h)

O out (mol/h) = 2 x F O_2 out (mol/h) + F Phenol solid out (mol/h) + F H_2O solid out (mol/h)

Balance: O in = O out

N-balance:

N in (mol/h) = 2 x F N_2 in (mol/h)

N out (mol/h) = 2 x F N_2 out (mol/h)

Balance: N in = N out

Table 9.4-2 presents the flow rates measured for the given co-feed experiment (PdGa, T=150°C, H_2/O_2=1.4) both for the gas species entering the reactor (dosage via mass flow controllers) and for the species exiting the reactor (determined via online µGC and offline GC-MS analysis). Due to the low benzene consumption observed in the experimental part, a precision of 1×10^{-5} mol/h is given for the calculated values. Please note that the precision of the determined rates for the different species deviates depending on the employed method. The precision of the entrance flows via the MFCs is ensured, for instance, at 1×10^{-2} mol/h, while the identification of the respective species, provided by the µGC and GC-MS, is reliable down to 1×10^{-9} mol/h. The total exit flow is measured with a precision of 1×10^{-3} mol/h as is the water rate, which is measured offline.

Table 9.4-2: Realized species balance for the co-feed hydroxylation experiment performed with the 5 µm PdGa-modified foil at 150°C at a H_2/O_2 ratio = 1.4.

Flows in (mol/h) – dosed via MFCs		Flows out (mol/h) – determined by µGC/GC-MS	
N_2 in	0.20006	N_2 out	0.20076
Benzene in	0.00516	Benzene out	0.00512
H_2 in	0.04156	H_2 out	0.03762
O_2 in	0.03000	O_2 out	0.02808
		Phenol out	1.123x10-7
		CO_2 out	5.555x10-7
		H_2O out	0.00377

It can be determined from Table 9.4-2 that the reactant consumption is very low, in particular for the C-species; this indicates the high stability and low reactivity of benzene in the gas phase under these experimental conditions. H- and O-species are more reactive; their consumption, however, is mainly related to the direct reaction of hydrogen and oxygen to water on the Pd-based catalyst. N-species are not reactive in the system. The slight deviation observed between the "in" and "out" values (an increase of approx. 0.4% was calculated) is probably due to the difference in quantification methods between the gases dosed in the reactor (MFCs) and those measured at the exit (µGC). Using the values of Table 9.4-2, one can calculate the global species rate in and out of the system using the above element balance. The results are given in Table 9.4-3. An important deviation (approx. > 5%) in the balance between the "in" and "out" values would indicate an incomplete balance in an element; this means that a reaction would have been omitted and that one or more reaction product(s) would not have been detected.

Table 9.4-3: Balance on the different elements for the experiment described in Table 9.4-2

Sum of C in (mol/h)	Sum of C out (mol/h)	Deviation from in value (%)
0.03098	0.03073	0.83
Sum of H in (mol/h)	**Sum of H out (mol/h)**	**Deviation from in value (%)**
0.11410	0.11353	0.50
Sum of O in (mol/h)	**Sum of O out (mol/h)**	**Deviation from in value (%)**
0.06000	0.05994	0.09
Sum of N in (mol/h)	**Sum of N out (mol/h)**	**Deviation from in value (%)**
0.20006	0.20076	0.34

Very low deviation values between the "in" and "out" elemental balances of the reactor are obtained, here there are all below 1%. The correlation in the balance for the C-, O- and H-products in the systems shows that all reactions, occurring in the system when performing the benzene hydroxylation with gas phase hydrogen and oxygen, have been considered.

9.4.2. 1.4 µm PdGa membrane

The reactor temperature was maintained at 150°C for all experiments. Some experiments were performed with membrane dosage mode, while others with all reactants in co-feed mode. In double-membrane dosage, the H_2 and O_2 gas partial pressures in the reaction channels were changed by adjusting the retentate pressures up to a value of 3 bar. In co-feed, the partial pressures were set by the entrance flows of the different gases. Double-membrane dosage experiments resulted, with the 1.4 µm thick PdGa layer, in hydrogen partial pressures up to 0.02 bar in the reactor, whereas the co-feed enabled higher hydrogen concentrations with P_{H2} up to 0.14 bar. Note that both oxygen and hydrogen partial pressures were varied from one experiment to the other. The results of this series of experiments are presented in Table 9.4-4.

Table 9.4-4: Parameters and results of the experiment series performed varying the partial pressures (via double-membrane dosage and co-feed) using the 1.4 µm thick PdGa-modified foil.

Series of exp.	Surface modifica-tion type	O_2 feed type	T (°C)	P_{H2} (bar)	H_2/ O_2	$X_{Benzene}$ (%)	Phenol rate $x10^{-8}$ (mol/h)	Phenol selectivity (%)	CO_2 rate $x10^{-7}$ (mol/h)	H_2O Rate $x10^{-3}$ (mol/h)
S11	1.4 µm PdGa	UF-TR.	150	0.007	1.3	0.004	1.3	26.2	2.2	1.2
S11	1.4 µm PdGa	UF-TR.	150	0.01	0.5	0.006	6.0	64.3	2.0	4.4
S11	1.4 µm PdGa	UF-TR.	150	0.02	0.5	0.004	7.1	64.0	2.4	4.3
S11	1.4 µm PdGa	co-feed	150	0.13	1.4	0.005	10.2	52.9	5.4	5.5

As expected, the phenol rate increases as the hydrogen partial pressures in the central channels increase up to the value permitted by the thinner PdGa layer, which is 0.02 bar. At this value, 64% of selectivity has been reached, which is more than that observed for the 5 µm layer. The phenol rate is also higher, here $7.1x10^{-8}$ mol/h was obtained. At P_{H2}=0.007 bar and H_2/O_2=1.4, the 5 µm layer featured a phenol rate of $1.6x10^{-8}$ mol/h with a selectivity of

26%. The 1.4 µm PdGa layer presented almost identical results with a rate of 1.3×10^{-8} mol/h and a phenol selectivity of 26.2%. A noticeable difference, however, was that the observed maximum for phenol selectivity was at a H_2/O_2 ratio of 0.5 in the case of this layer thickness; the peak was obtained for a ratio of 1.4 with 5 µm PdGa. One, however, has to be take into consideration that the 5 µm PdGa membrane was tested in co-feed mode, whereas the 1.4 µm membrane was basically used in membrane-dosage and the gas concentration profiles along the reactor are thus different.

The co-feed experiment where a higher P_{H2} was obtained (0.14 bar) logically resulted in higher product rates with 1.0×10^{-7} mol/h for phenol, 5.4×10^{-7} mol/h for CO_2 and 5.5×10^{-3} mol/h for water respectively. Quasi identical results were obtained in co-feed with the 5 µm PdGa layer under the same conditions: 1.1×10^{-7} mol/h for phenol, 5.5×10^{-7} mol/h for CO_2 and 3.8×10^{-3} mol/h for water respectively. The same surface state presented similar results as hydrogen was dosed at the reactor entrance in both configurations.

9.4.3. 0.4 µm PdGa membrane

Experiments using the 0.4 µm membrane were conducted on the same principle as for the 1.4 µm PdGa-based membrane. The reactor temperature was also kept constant at 150°C for all experiments. Most experiments were realized using membrane dosage, while one was in co-feed mode. Double-membrane dosage experiments with the thinner PdGa layer presented a hydrogen partial pressure range in the reaction channels up to 0.05 bar. The 0.4 µm PdGa layer technically delivers higher hydrogen concentrations than the 1.4 µm layer, it was still possible to cover a satisfactory partial pressure range using both membranes. The results of the series of experiments are shown in Table 9.4-5.

Table 9.4-5: Parameters and results of the experiment series performed varying the partial pressures (via double-membrane dosage and co-feed) using the 0.4 µm thick PdGa-modified foil.

Series of exp.	Surface modifica-tion type	O_2 feed type	T (°C)	P_{H2} (bar)	H_2/O_2	$X_{Benzene}$ (%)	Phenol rate $x10^{-8}$ (mol/h)	Phenol selectivity (%)	CO_2 rate $x10^{-7}$ (mol/h)	H_2O Rate $x10^{-3}$ (mol/h)
S12	0.4 µm PdGa	UF-TR.	150	0.007	1.4	0.004	1.7	29.0	2.5	0.8
S12	0.4 µm PdGa	UF-TR.	150	0.05	0.5	0.004	9.0	63.5	3.1	4.7
S12	0.4 µm PdGa	UF-TR.	150	0.05	1.4	0.002	6.8	49.3	4.2	4.7
S12	0.4 µm PdGa	co-feed	150	0.14	1.4	0.004	9.6	52.6	5.2	3.6

The experiments realized with this membrane, at a higher hydrogen partial pressure in the system (0.05 bar as opposed to 0.02 bar), featured very similar results when compared to the 1.4 µm layer. The maximum of phenol selectivity, 63.5 %, was also measured at $H_2/O_2=0.5$. The average phenol rate, under these conditions, was $7.9x10^{-8}$ mol/h, while the average CO_2 and water rates were $3.6x10^{-7}$ mol/h and $4.7x10^{-3}$ mol/h respectively. For the 1.4 µm layer, these values were $6.5 x10^{-8}$ mol/h, $2.2 x10^{-7}$ mol/h and $4.3 x10^{-3}$ mol/h respectively.

The experiment in co-feed showed the highest phenol and CO_2 rates of the series of experiments but the water rate did not follow the same increase. The water rate even slightly decreased from $4.7x10^{-3}$ mol/h to $3.6x10^{-3}$ mol/h, in spite of the higher oxygen and hydrogen partial pressures. This could also be attributed to the fact that the PdGa surface limits, to some extent, the CO_2 and water rates due to the particular surface arrangement of this catalyst.

A comparison between the 3 PdGa membranes was facilitated by performing the experiments in double-membrane dosage at very low hydrogen partial pressures in order to reproduce the process conditions obtained with the 5 µm PdGa membrane. In this connection, both the 0.4 µm and the 1.4 µm PdGa membranes were used, a reaction temperature of 150°C was set and the hydrogen retentate pressure was adjusted, in each case, so that the partial pressure inside the reactor did not exceed 0.007 bar. The H_2/O_2 ratio was maintained at 1.4. The benzene partial pressure was 0.04 bar for all experiments. Fig. 9.4-1 shows the comparison between the product rates over the 3 PdGa surfaces, including the previous results obtained over 5 µm PdGa (see Table 9.4-4).

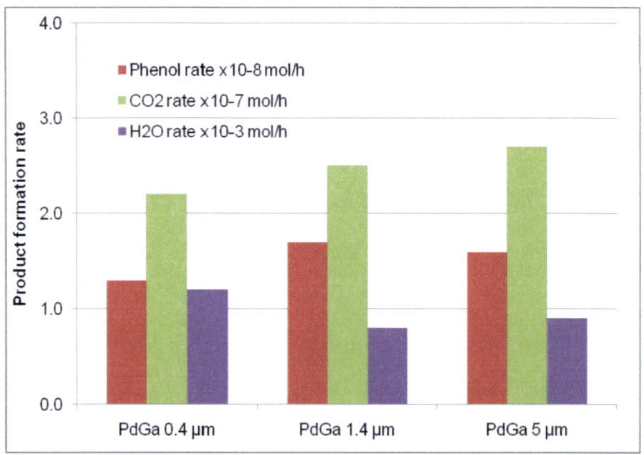

Fig. 9.4-1. Performance comparison of the different PdGa sputtered layers in double-membrane operation mode (T: 150°C; H_2/O_2: 1.4; P_{O2} = 0.007 bar).

No significant difference is apparent between the systems; the phenol, CO_2 and H_2O rates are stable at values of around 1.5×10^{-8} mol/h, 2.5×10^{-7} mol/h and 1.5×10^{-3} mol/h respectively. Only small variations are discernable between the layers; for instance, the maximal phenol rate was measured over 1.4 µm PdGa, the minimal CO_2 rate over 0.4 µm PdGa, while the minimal water rate was measured over 1.4 µm PdGa. These variations, however, do not exceed 15% of the measured average rate and are therefore considered stable. The thickness of the catalytic layer influences the hydrogen permeation through the membrane and imposes a limit in terms of the attainable partial pressure for the reaction; the layer thickness, however, does not intrinsically influence the product formation rates (phenol and byproducts). In this sense, we can consider the 3 PdCu foils, used as the support, to be homogeneously covered

– all over the surface – by a PdGa layer, even in the case of the 400 nm thick sputtered film.

9.5. Comparison of the performance of the different catalytic systems

9.5-1. Double-membrane dosage operation

Data are available from experiments in double-membrane mode (PdCu, $Pd_{90}Au_{10}$, $Pd_{90}Au_{10}$ on V_2O_5 and $Pd_{50}Ga_{50}$, as thinner layers were produced) and in co-feed mode (PdGa and $Pd_{90}Au_{10}$) for the comparison of the catalytic properties of the different layers. The reactor temperature was kept constant at 150°C, while the benzene partial pressure in the feed was set to 0.04 bar for all experiments. For those with hydrogen membrane dosage, the H_2/O_2 ratio was kept constant at a value of 1.4 in order to enable a comparison of the performance of the different catalytic layers. A narrow range of hydrogen and oxygen partial pressures in the reactor outlet was striven for, i.e. P_{H2} = 0.05 - 0.08 bar and P_{O2} by adjusting the retentate side gas pressures.

Fig. 9.5-1 compares the phenol formation rate, phenol selectivity and water formation rate detected in double-membrane dosage in the 4 systems: Pd on PdCu, $Pd_{90}Au_{10}$ and $Pd_{90}Au_{10}$ on V_2O_5.

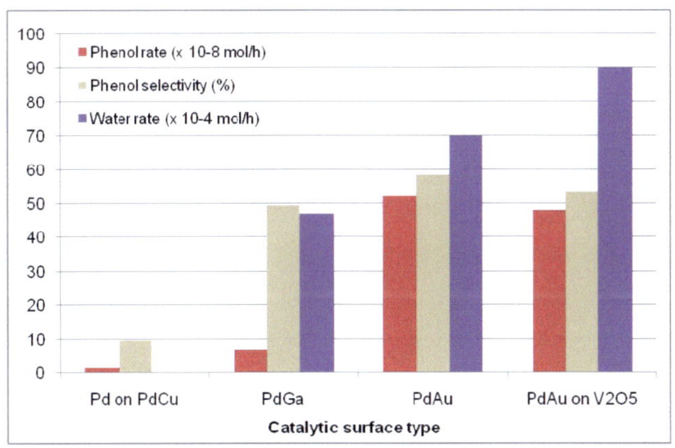

Fig. 9.5-1. Performance comparison of different catalytic layers in double-membrane operation (T: 150°C; H_2/O_2: 1.4; P_{O2} = 0.05 bar).

It can be seen that both PdAu systems clearly outperform the reference Pd on PdCu system in terms of phenol productivity and phenol selectivity. The best performance of the $Pd_{90}Au_{10}$ layer in double-membrane mode at 150°C obtained so far was reached at a slightly reduced H_2/O_2 ratio of 1.0 (i.e. $P_{H2} = P_{O2} = 0.05$ bar). The phenol formation rate reached 7.2×10^{-7} mol/h at a phenol selectivity of 67.6 % (carbon-based); benzene conversion was 0.023 % under these conditions. Positive performances were also obtained for V_2O_5/PdAu (phenol rate of 5.4×10^{-7} mol/h and selectivity of 59.8%); they were, however, not as high as they should have been on an enhanced catalyst surface. By comparing the results for the two PdAu systems, it is clear that the aim of increasing the catalytic surface area by introducing a discontinuous V_2O_5 support layer has not been achieved. This has already been ascertained from the EDX results, which suggested a continuous film of $Pd_{90}Au_{10}$, rather than discrete particles on the V_2O_5 support as had originally been anticipated.

The PdGa system also outperforms the Pd/PdCu system but more in terms of phenol selectivity as the rate is "only" improved by a factor 5. The water formation rate for the Pd on PdCu system is not known due to an initial limitation in the experimental setup. While the carbon-based phenol selectivity for the PdAu systems is acceptable with values around 50 %, the hydrogen-based phenol selectivity is clearly not. The water formation rate is higher by 4 orders of magnitude, resulting in a hydrogen-based phenol selectivity in the range of 0.01 %.

The catalytic systems showing the highest phenol rates are first of all PdAu followed by V_2O_5/PdAu, with rates approx. 7 times higher than those for PdGa. PdAu shows the best C-selectivity to phenol, followed by V_2O_5/PdAu, the selectivity measured over PdGa is, however not far behind (49% for PdGa as opposed to 53% for V_2O_5/PdAu and 67.6% for PdAu).

In terms of the formation of by-products, the maximum with regard to the restriction of CO_2 and water formation, under the same process conditions, is realized with the PdGa system. The CO_2 rate on PdGa is 4 times lower than on PdAu, due to the geometric distribution of the Pd active centers on the surface, which influence the benzene adsorption. The weaker adsorption of benzene, theoretically achieved over PdGa and the limited amount of Pd in the catalyst compared to the 2 other systems, are probably responsible for the lower generation of byproducts in this case. The rate of the other side-reaction, the water formation, is also reduced by a factor 2 on the PdGa surface due to the same effect. In this sense, PdGa is the catalyst with the best potential for limiting the influence of side-reactions. Its use could, therefore, be interesting in the study of other reaction systems involving hydrogen where a competition between reactions exists.

9.5-2. Co-feed operation

To overcome the initial permeation limitation with 5 µm PdGa, experiments were carried out with this system where hydrogen and oxygen were introduced together with benzene in the feed (co-feed). A comparison between $Pd_{90}Au_{10}$ and PdGa can be made based on the data from the co-feed experiments with the 5 µm $Pd_{50}Ga_{50}$ layer at different hydrogen partial pressures and the results of the $Pd_{90}Au_{10}$ system under the same conditions. $Pd_{90}Au_{10}$ results interpolated from the data can be seen in Figs. 9.2-4, 9.2-5 and 9.2-6. Table 9.5-1 shows the results, which reveal a superior catalytic performance of PdGa in terms of lower H_2O and CO_2 formation rates, in particular at the lower H_2/O_2 ratio of 1.4. Moreover, the sensitivity of the performance towards the H_2/O_2 ratio (or P_{H2}, in this case) appears to be higher for PdGa than for $Pd_{90}Au_{10}$, which might be linked to the fact that PdGa, contrary to $Pd_{90}Au_{10}$, does not readily absorb hydrogen.

Table 9.5-1. Comparison of the catalytic performance of 1 µm $Pd_{90}Au_{10}$ and 5 µm PdGa in hydrogen co-feed mode (T: 150°C, P: 1 bar; benzene partial pressure: 0.04 bar, P_{O2} = 0.1 bar, P_{H2} varied).

Catalytic layer	$Pd_{90}Au_{10}$	PdGa
H_2/O_2 ratio = 1.4 P_{O2} = 0.1 bar	Phenol : 4.0×10^{-8} mol/h CO_2 : 2.7×10^{-6} mol/h H_2O : 6.4×10^{-3} mol/h	Phenol : 1.1×10^{-7} mol/h CO_2 : 1.2×10^{-7} mol/h H_2O : 3.8×10^{-3} mol/h
H_2/O_2 ratio = 2.0 P_{O2} = 0.1 bar	Phenol : 3.6×10^{-8} mol/h CO_2 : 2.7×10^{-6} mol/h H_2O : 6.7×10^{-3} mol/h	Phenol : 3.0×10^{-8} mol/h CO_2 : 1.8×10^{-7} mol/h H_2O : 4.5×10^{-3} mol/h

A phenol rate above 1×10^{-7} mol/h was interestingly measured for an optimal gas mixture over PdGa in co-feed mode. A similar observation to a double-membrane dosage can be realized for a H_2/O_2 ratio of 2.0: the PdAu system shows a slightly higher phenol rate than PdGa (however, not as high as in the double-membrane dosage mode, a factor of only 1.2 is obtained here) as well as higher side-reaction rates. The lower rate obtained over PdAu in co-feed mode, in comparison to the double-membrane mode, affects the phenol C-selectivity. A selectivity of only about 5% was calculated, whereas the PdGa surface shows a higher phenol selectivity, which is ten times higher.

PdGa again shows interesting behavior in terms of limiting the CO_2 and water rates, in comparison to PdAu, where they are produced in similar amounts to the membrane dosage experiments. On PdGa in co-feed mode, the CO_2 and water rates are 15 times and 1.5 times lower than on PdAu respectively, which emphasizes the benefit of the surface properties of this catalyst towards side-reactions involving hydrogen in addition to hydroxylation.

10. Modeling reaction kinetics

10.1. Langmuir-Hinshelwood model

The experimental part, which involved performing benzene hydroxylation with the PdAu modified membrane at 150°C, yielded information on the product rates under different reaction conditions (benzene, hydrogen, oxygen partial pressures). These product rates (phenol, CO_2, H_2O) depend on the state of the catalytic surface during the process, in particular, the proportion and state of the species adsorbed on the PdAu membrane surface (absorbed benzene, dissociated hydrogen species and interaction with the adsorbed oxygen molecule). A kinetic model, which is based on the adsorption of the available gaseous species and which takes into account surface reactions to describe the formation rate of the different products in the system, is proposed in this chapter. Elementary steps involving all species introduced into the reactor are described. These steps aim to supply the reactive elements, which subsequently lead to the formation of the different products via surface reactions. Their description is based on the very well-known interaction between hydrogen and palladium alloys, as the catalytic surfaces, whereby the dissociative adsorption of hydrogen is the initial step. Assumptions on interactions and favored reactions are also made in order to orient the system description towards the experimental observations in terms of the type and rates of the products formed in the reactor. The ability of the proposed model to reproduce the obtained experimental data should allow for a validation of the previously made assumptions.

10.1.1. Theoretical approach considering adsorption/desorption of the reactive species

The following approach describes, using fundamental reaction equations, the absorption, desorption and surface reaction phenomena in the adsorbed state, which occur in the reactor according to the Langmuir-Hinshelwood mechanism. Johansson et al. used such a model to describe the water formation over polycrystalline palladium and platinum catalysts [71]. However, the experimental conditions used in their work differ a lot from those in the double-membrane reactor. Indeed, Johansson et al. studied the H_2/O_2 reaction by performing laser-induced fluorescence and microcalorimetry on palladium at a temperature of 1300 K; they supplied coefficients for the different involved steps (adsorption and surface reaction). Other authors have also studied water reaction kinetics over Pd surfaces [73-74]; the data supplied there, however, do not allow for a calculation of the coefficients in a Langmuir-Hinshelwood model. Even though the extrapolation of Johansson's study of the water reaction to a temperature of 423 K and a $Pd_{90}Au_{10}$ wt.% surface appears drastic, this strategy was chosen to develop a kinetics model with the aim of reproducing the behavior observed in the double-membrane reactor. Every elementary step of the Langmuir-Hinshelwood mechanism is listed and described below with its related reaction assumptions.

The initial hypothesis assumes that there is a unique type of adsorption site for all species on the catalyst surface. The adsorption of the different gas species can occur anywhere on the membrane surface. The adsorption and desorption steps are also considered to be in equilibrium, much faster than the reaction steps. It is assumed that the surface coverage of adsorbed species depends on the gas phase composition at the equilibrium. It is also assumed that the benzene molecule adsorbs 1 Pd center. It was shown in several studies that benzene is usually adsorbed on 3 neighboring Pd centers [67-69]; this, however, would lead to a very complex an expression for the fraction of free sites with the partial pressures and for this reason had to be simplified.

1- Dissociative adsorption of hydrogen on 2 neighboring free centers [71]:

$$H_2 + 2\,[\,] \leftrightarrow 2\,[H] \qquad\qquad\qquad [R1]$$

We assume here that adsorption and desorption are in equilibrium, i.e. much faster than other reactions, so that:

$$k1_{ads}.P_{H2}.\Theta_f^2 = k1_{des}.\Theta_H^2$$

setting $K_{H2} = k1_{ads}/k1_{des}$ we obtain

$$K_{H2}.\, P_{H2}.\Theta_f^2 = \Theta_H^2$$

$$\Rightarrow \Theta_H = \sqrt{K_{H2}P_{H2}}\,.\,\Theta_f \qquad\qquad \text{Eq. (10.1-1)}$$

2- Adsorption of oxygen on 2 neighboring free centers.

We assume that oxygen dissociates on the catalyst as well, so that:

$$O_2 + 2\,[\,] \leftrightarrow 2\,[O] \qquad\qquad [R2]$$

Similarly as for hydrogen adsorption, if adsorption and desorption are in equilibrium:

$$k2_{ads}.P_{O2}.\Theta_f^2 = k2_{des}.\Theta_O^2$$

with $K_{O2} = k2_{ads}/k2_{des}$:

$$K_{O2}.\, P_{O2}.\Theta_f^2 = \Theta_O^2$$

$$\Rightarrow \Theta_O = \sqrt{K_{O2}P_{O2}}\,.\,\Theta_f \qquad\qquad \text{Eq. (10.1-2)}$$

3- Adsorption of benzene on 1 free center:

$$B + [\,] \leftrightarrow [B] \qquad\qquad [R3]$$

One assumes again that adsorption and desorption are in equilibrium, i.e. much faster than other reactions, so that:

$$k3_{ads}.P_B.\Theta_f = k3_{des}.\Theta_B$$

with $K_B = k3_{ads}/k3_{des}$:

$$\Rightarrow \Theta_B = K_B.P_B.\Theta_f \qquad\qquad \text{Eq. (10.1-3)}$$

4- Reaction between adsorbed hydrogen and oxygen species to produce surface hydroxyl species [71]:

$$[H] + [O] \leftrightarrow [OH] + [\,] \qquad\qquad [R4]$$

The reaction kinetics can be described by:

$$r4 = k4^{+}.\Theta_H.\Theta_O - k4^{-}.\Theta_{OH}.\Theta_f \qquad\qquad \text{Eq. (10.1-4)}$$

It is expected that the direct production of phenol from benzene occurs via the reaction between the adsorbed benzene and OH radicals on the surface of the Pd-based membrane [24]. Reaction R4 is taken to be reversible and much quicker than any of the surface reactions; [OH] species are thus permanently present on the catalyst surface and can be utilized to hydroxylate the benzene. The following applies in the case of reaction equilibrium of reaction R4:

$$k4^{+}.\Theta_H.\Theta_O = k4^{-}.\Theta_{OH}.\Theta_f$$

with $K_4 = k4^{+}/k4^{-}$:

$$\Theta_{OH} = K_4.\ \Theta_H.\Theta_O/\Theta_f$$

Using the expressions of Θ_H and Θ_O from R1 and R2:

$$\Theta_{OH} = K_4.\ \sqrt{K_{H2}P_{H2}}\ .\sqrt{K_{O2}P_{O2}}\ .\ \Theta_f \qquad\qquad \text{Eq. (10.1-5)}$$

5- Reaction between adsorbed hydroxyl and hydrogen species to produce water:

$$[OH] + [H] \leftrightarrow [H_2O] + [\,] \qquad\qquad [R5]$$

The water formation kinetics can be expressed by:

$$r5 = k5^{+}.\Theta_{OH}.\Theta_H - k5^{-}.\Theta_{H2O}.\Theta_f \qquad\qquad \text{Eq. (10.1-7)}$$

It is to be expected from a thermodynamic aspect that the water formation, according to R5, is an irreversible process so that Eq. (10.1-7) can be further simplified (see expression of the water formation kinetics further on).

6- Reaction between 2 adsorbed hydroxyl species to produce water:

$$[OH] + [OH] \leftrightarrow [H_2O] + [O] \qquad\qquad [R6]$$

Here, the water formation kinetics is given by:

$$r6 = k6^+.\Theta_{OH}^2 - k6^-.\Theta_{H2O}.\Theta_O \qquad\qquad Eq.\ (10.1\text{-}8)$$

The question as to whether water would react with adsorbed oxygen species on the catalyst to form two surface hydroxyl species, or whether the water molecule would be too stable to dissociate is more difficult to assess here. Both cases can be considered in the expression of the water formation kinetics.

7- Reaction between benzene and adsorbed hydroxyl radical for the production of phenol:

$$[B] + [OH] \leftrightarrow [Ph] + [H] \qquad\qquad [R7]$$

This reaction would be the only one leading to the formation of phenol from benzene. The reaction kinetics can be expressed by:

$$r7 = k7^+.\Theta_B.\Theta_{OH} - k7^-.\Theta_{Ph}.\Theta_H \qquad\qquad Eq.\ (10.1\text{-}9)$$

The adsorption steps of the products formed have to be described:

8- Adsorption of water on a free center:

$$H_2O + [\ ] \leftrightarrow [H_2O] \qquad\qquad [R8]$$

Again, assuming that adsorption and desorption phases are in equilibrium, i.e. much faster than other reactions:

$$k8_{ads}.P_{H2O}.\Theta_f = k8_{des}.\Theta_{H2O}$$

with $K_{H2O} = k8_{ads}/k8_{des}$:

$$\Theta_{H2O} = K_{H2O}.P_{H2O}.\Theta_f \qquad\qquad Eq.\ (10.1\text{-}10)$$

9- Adsorption of phenol on 1 free center (assumption of same adsorption behavior as for benzene):

$$Ph + [\,] \leftrightarrow [Ph] \qquad\qquad [R9]$$

Similarly, with adsorption and desorption in equilibrium we obtain:

$$k9_{ads}.P_{Ph}.\Theta_f = k9_{des}.\Theta_{Ph}$$

with $K_{Ph} = k9_{ads}/k9_{des}$:

$$\Rightarrow \Theta_{Ph} = K_{Ph}.P_{Ph}.\Theta_f \qquad\qquad \text{Eq. (10.1-11)}$$

A mechanism, which leads to the formation of carbon dioxide in the reactor, must be described at this stage. Very low partial orders on hydrogen and oxygen were measured for the CO_2 formation rate during the experiments. Based on Fig. 9.2-5, an exponent of 0.31 applies for the oxygen partial pressure for the description of the CO_2 rate; in the experiments, the CO_2 formation rate was shown to be independent of the hydrogen partial pressure (exponent null). In stoichiometrical terms, the total oxidation of benzene to CO_2 requires either 6 molecules of O_2 for one molecule of benzene for the reaction in non-adsorbed state, or a reaction between [B] and 12 [O] in adsorbed conditions. It cannot be definitely ascertained whether surface or gas phase oxygen is involved in the CO_2.

10- Reaction between adsorbed benzene and adsorbed oxygen to form CO_2. Benzene is adsorbed on 1 free center.

The formation of CO_2 from benzene requires, in stoichiometrical terms, the presence of 12 [O] sites on the surface surrounding the adsorbed benzene molecule. The following equation is obtained where this configuration is assumed possible:

$$[B] + 12\,[O] + \leftrightarrow 6\,[CO_2] + 6\,[H] + 1\,[\,] \qquad\qquad [R10]$$

$$r10 = k10^{+}.\Theta_B.\Theta_O^{12} - k10^{-}.\Theta_{CO2}^{6}.\Theta_H^{6}.\Theta_f \qquad\qquad \text{Eq. (10.1-12)}$$

11- Reaction between adsorbed phenol and adsorbed oxygen for the formation of CO_2. Assumption: phenol adsorbed on 1 free site, as for benzene:

A similar process could apply to the phenol oxidation mechanism, as for benzene:

$$[Ph] + 12\,[O] + \leftrightarrow 6\,[CO_2] + 5\,[H] + [OH] + [\,] \qquad [R11]$$

The related formation kinetics would be:

$$r11 = k11^{+}.\Theta_{Ph}.\Theta_{O}^{12} - k11^{-}.\Theta_{CO2}^{6}.\Theta_{H}^{5}.\Theta_{OH}.\Theta_{f} \qquad \text{Eq. (10.1-13)}$$

12- Adsorption of CO_2 on a free center:

$$CO_2 + [\,] \leftrightarrow [CO_2] \qquad [R12]$$

Similarly:

$$k12_{ads}.P_{CO2}.\Theta_{f} = k12_{des}.\Theta_{CO2}$$

with $K_{CO2} = k12_{ads}/k12_{des}$:

$$\Theta_{CO2} = K_{CO2}.P_{CO2}.\Theta_{f} \qquad \text{Eq. (10.1-14)}$$

Assuming that there is only one type of adsorption site, the balance on the free centers on the PdAu surface would result in:

$$\Theta_{H} + \Theta_{O} + \Theta_{B} + \Theta_{OH} + \Theta_{H2O} + \Theta_{Ph} + \Theta_{CO2} + \Theta_{f} = 1 \qquad \text{Eq. (10.1-15)}$$

When combining the expressions of R1, R2, R3, R4, R8, R9 and R12 and introducing them into the Eq. (10.1-15), an expression, which is only a function of Θ_{f} is found:

$$\Theta_{f}.(1 + \sqrt{K_{H2}P_{H2}} + \sqrt{K_{O2}P_{O2}} + K_{4}.\sqrt{K_{O2}P_{O2}}\cdot\sqrt{K_{H2}P_{H2}} + K_{B}.P_{B} + K_{Ph}.P_{Ph} + K_{H2O}.P_{H2O} + K_{CO2}.P_{CO2}) = 1$$

$$\Theta_f = \frac{1}{1+\sqrt{K_{H2}P_{H2}}+\sqrt{K_{O2}P_{O2}}+K_4\cdot\sqrt{K_{O2}P_{O2}}\cdot\sqrt{K_{H2}P_{H2}}+K_B P_B+K_{Ph}P_{Ph}+K_{H2O}P_{H2O}+K_{CO2}P_{CO2}}$$

Eq. (10.1-16)

From this point on, the different product formation rates can be modeled using adsorption/desorption equilibrium constants and the kinetic coefficients of the reactions.

Expression of the water formation kinetics:

According to the formulated mechanism, water is produced by the reactions R5 and R6. During the experiments, it was observed that water is by far the main product in the system due to direct hydrogen oxidation, when H_2 and O_2 are dosed into the reactor. So that:

$rH_2O = r5 + r6$

$rH_2O = (k5^+.\Theta_{OH}.\Theta_H - k5^-.\Theta_{H2O}.\Theta_f) + (k6^+.\Theta_{OH}{}^2 - k6^-.\Theta_{H2O}.\Theta_O)$

At this point, one can consider 2 options. Assuming that R5 is an irreversible process, R6 can be seen either as reversible or irreversible.

If one considers the reactions R5 and R6 irreversible, one can simplify the water formation kinetics to the following expression:

Case A: R6 is irreversible:

$rH_2O = k5^+.\Theta_{OH}.\Theta_H + k6^+.\Theta_{OH}{}^2$ Eq. (10.1-17)

which gives:

$$rH_2O = \frac{k_{5+}.K_4\cdot\sqrt{K_{O2}}.K_{H2}\cdot\sqrt{P_{O2}}.P_{H2}+k_{6+}.K_4{}^2.K_{O2}.K_{H2}.P_{O2}.P_{H2}}{\left(1+\sqrt{K_{H2}P_{H2}}+\sqrt{K_{O2}P_{O2}}+K_4\cdot\sqrt{K_{O2}P_{O2}}\cdot\sqrt{K_{H2}P_{H2}}+K_B P_B+K_{Ph}P_{Ph}+K_{H2O}P_{H2O}+K_{CO2}P_{CO2}\right)^2}$$

Eq. (10.1-18)

Or:

Case B: R6 is reversible:

$$rH_2O = k5^+.\Theta_{OH}.\Theta_H + k6^+.\Theta_{OH}^2 - k6^-.\Theta_{H2O}.\Theta_O \qquad\qquad \text{Eq. (10.1-19)}$$

$$rH_2O = \frac{k_{5+}.K_4.\sqrt{K_{O2}}.K_{H2}.\sqrt{P_{O2}}.P_{H2} + k_{6+}.K_4^2.K_{O2}.K_{H2}.P_{O2}.P_{H2} - k_{6-}.\sqrt{K_{O2}}.K_{H2O}.\sqrt{P_{O2}}.P_{H2O}}{\left(1 + \sqrt{K_{H2}P_{H2}} + \sqrt{K_{O2}P_{O2}} + K_4.\sqrt{K_{O2}P_{O2}}.\sqrt{K_{H2}P_{H2}} + K_B P_B + K_{Ph}P_{Ph} + K_{H2O}P_{H2O} + K_{CO2}P_{CO2}\right)^2}$$

Eq. (10.1-20)

Both cases will be discussed further on.

Expression of the phenol formation kinetics:

The total phenol production rate can be expressed by the rate of the R7 reaction (benzene hydroxylation) minus the rate of the R11 reaction (phenol consumption due to oxidation to CO_2), which gives:

$$rPh = r7 - r11$$

$$rPh = (k7^+.\Theta_B.\Theta_{OH} - k7^-.\Theta_{Ph}.\Theta_H) - (k11^+.\Theta_{Ph}.\Theta_O^{12} - k11^-.\Theta_{CO2}^6.\Theta_H^5.\Theta_{OH})$$

Reaction R11 can be taken to be irreversible; it is improbable that CO_2 will react with surface species to form phenol. On the other hand, R7 could be reversible, 2 options must thus be considered.

Case A: R7 is irreversible:

The maximum in the phenol rate would be then explained by an increasing phenol oxidation and not a reversibility of the phenol reaction:

$$rPh = k7^+.\Theta_B.\Theta_{OH} - k11^+.\Theta_{Ph}.\Theta_O^{12} \qquad\qquad \text{Eq. (10.1-21)}$$

This gives:

$$rPh = \frac{k_{7+}.K_B.K_4.\sqrt{K_{O2}}.\sqrt{K_{H2}}.P_B.\sqrt{P_{O2}}.\sqrt{P_{H2}}}{\left(1+\sqrt{K_{H2}P_{H2}}+\sqrt{K_{O2}P_{O2}}+K_4.\sqrt{K_{O2}P_{O2}}.\sqrt{K_{H2}P_{H2}}+K_B P_B+K_{Ph}P_{Ph}+K_{H2O}P_{H2O}+K_{CO2}P_{CO2}\right)^2}$$

$$- \frac{k_{11+}.K_{Ph}.K_{O2}{}^6.P_{Ph}.P_{O2}{}^6}{\left(1+\sqrt{K_{H2}P_{H2}}+\sqrt{K_{O2}P_{O2}}+K_4.\sqrt{K_{O2}P_{O2}}.\sqrt{K_{H2}P_{H2}}+K_B P_B+K_{Ph}P_{Ph}+K_{H2O}P_{H2O}+K_{CO2}P_{CO2}\right)^{13}}$$

Eq. (10.1-22)

Or:

Case B: R7 is reversible:

Then: $rPh = k7^+.\Theta_B.\Theta_{OH} - k7^-.\Theta_{Ph}.\Theta_H - k11^+.\Theta_{Ph}.\Theta_O{}^{12}$ Eq. (10.1-23)

$$rPh = \frac{k_{7+}.K_B.K_4.\sqrt{K_{O2}}.\sqrt{K_{H2}}.P_B.\sqrt{P_{O2}}.\sqrt{P_{H2}} - k_{7-}.K_{Ph}.\sqrt{K_{H2}}.P_{Ph}.\sqrt{P_{H2}}}{\left(1+\sqrt{K_{H2}P_{H2}}+\sqrt{K_{O2}P_{O2}}+K_4.\sqrt{K_{O2}P_{O2}}.\sqrt{K_{H2}P_{H2}}+K_B P_B+K_{Ph}P_{Ph}+K_{H2O}P_{H2O}+K_{CO2}P_{CO2}\right)^2}$$

$$- \frac{k_{11+}.K_{Ph}.K_{O2}{}^6.P_{Ph}.P_{O2}{}^6}{\left(1+\sqrt{K_{H2}P_{H2}}+\sqrt{K_{O2}P_{O2}}+K_4.\sqrt{K_{O2}P_{O2}}.\sqrt{K_{H2}P_{H2}}+K_B P_B+K_{Ph}P_{Ph}+K_{H2O}P_{H2O}+K_{CO2}P_{CO2}\right)^{13}}$$

Eq. (10.1-24)

Expression of the CO_2 formation kinetics:

CO_2 formation kinetics is derived from the sum of the benzene oxidation and phenol oxidation, which are the only C-containing species involved here, so that:

$rCO_2 = r10 + r11$

Using the rates expressed by Eq. (10.1-12) and Eq. (10.1-13), and assuming the reactions are irreversible, then:

$$rCO_2 = \frac{k_{10+}.K_B.K_{O2}{}^6.P_B.P_{O2}{}^6}{\left(1+\sqrt{K_{H2}P_{H2}}+\sqrt{K_{O2}P_{O2}}+K_4.\sqrt{K_{O2}P_{O2}}.\sqrt{K_{H2}P_{H2}}+K_B P_B+K_{Ph}P_{Ph}+K_{H2O}P_{H2O}+K_{CO2}P_{CO2}\right)^{13}}$$

$$+ \frac{k_{11+}.K_{Ph}.K_{O2}{}^6.P_{Ph}.P_{O2}{}^6}{\left(1+\sqrt{K_{H2}P_{H2}}+\sqrt{K_{O2}P_{O2}}+K_4.\sqrt{K_{O2}P_{O2}}.\sqrt{K_{H2}P_{H2}}+K_B P_B+K_{Ph}P_{Ph}+K_{H2O}P_{H2O}+K_{CO2}P_{CO2}\right)^{13}}$$

Eq. (10.1-25)

A discussion on the rate expressions based on Eq. (10.1-18), Eq. (10.1-20), Eq. (10.1-22), Eq. (10.1-24) and Eq. (10.1-25) allows for an analysis of their ability to reproduce (or not) the profiles, which were obtained from the experiments and are described in Chapter 9 in the case of the PdAu catalytic surface.

10.2. Model vs. experimental data: case of PdAu

10.2.1. Discussion on the expression of the proposed product rate laws

Water rate:

The modeling of the water formation rate through the successive hypotheses of surface reaction mechanisms resulted in the expressions found in Eq. (10.1-18) and Eq. (10.1-20). Both cases have to be analyzed depending on whether R6 is reversible or not.

Case A: R6 irreversible

Eq. (10.1-18) is composed of the addition of two terms related to P_{O2} and P_{H2} with a common denominator, which is also related to P_{O2} and P_{H2}. The denominator has the form $(1+X)^2$ where X is the sum of terms containing the partial pressures. Different partial orders on P_{O2} and P_{H2} are obtained depending on whether X is "big" or negligible with respect to 1.

If the term containing the partial pressures at the denominator (namely

$$\sqrt{K_{H2}P_{H2}} + \sqrt{K_{O2}P_{O2}} + K_4 \cdot \sqrt{K_{O2}P_{O2}} \cdot \sqrt{K_{H2}P_{H2}} + K_B P_B + K_{Ph}P_{Ph} + K_{H2O}P_{H2O} + K_{CO2}P_{CO2})$$ is

much bigger than 1, then the denominator would show a partial order of 1 for O_2 and 1 for H_2. Due to the fact that the numerator is at the partial order ranging from 0.5 to 1 for O_2 and 1 for H_2 respectively, this results in an overall partial order, which ranges between -0.5 and 0 for O_2 and a partial order of 0 for H_2.

On the other hand, if the term containing the partial pressures at the denominator is negligible with respect to 1, then the denominator would be approximated to 1 and only the numerator would remain in the expression of the water rate. A partial order ranging from 0.5 to 1 for O_2 and at a partial order 1 for H_2 would then be obtained.

In both cases, the partial order on hydrogen is superior to that for oxygen in the model.

Experimentally, partial orders of 0.023 for O_2 and 0.122 for H_2 were observed. This is in agreement with the range of partial orders provided by the model. Moreover, both experimental partial orders have intermediary values between the two above-mentioned extreme cases (partial order being between -0.5 and 1 on O_2 and between 0 and 1 on H_2). With appropriate K values, the sum of the terms containing the partial pressures at the denominator can be "close to 1" in value, and can thus reproduce the intermediary partial order that was observed experimentally. When a qualitative analysis is carried out, it becomes apparent that Eq. (10.1-18) is able to reproduce the experimentally observed trend and can thus fit the experimental data.

Case B: R6 reversible

Eq. (10.1-20) is composed of three terms related to P_{O2} and P_{H2} with the same common denominator, which is also related to P_{O2} and P_{H2}. Both the numerator and denominator show the same partial orders for oxygen and hydrogen as in the previous case. The same conclusion, on the ability of the model to reproduce the influence of the oxygen and hydrogen partial pressures on the experimentally observed water rate, can thus be made. As a notable change, the coefficient in front of $P_{O2}^{0.5}$ at the numerator will be smaller than in case A; this would lower the water rate in comparison to case A but would not modify the shape of the modeled profile.

<u>Phenol rate:</u>

The expression of the proposed phenol rate can be found in Eq. (10.1-22) and Eq. (10.1-24).

Case A: R7 irreversible

The phenol rate is composed of a (positive) phenol formation term and a (negative) phenol consumption term, which have the same denominator but at different exponents. Experimentally, a maximum in the phenol rate was observed where the oxygen and hydrogen partial pressures were both varied. The rate decrease after reaching the maximum can be explained by the reaction conditions, which lead to an acceleration of the phenol oxidation to CO_2 in comparison to the phenol formation from benzene. In this connection, an increasing influence of the negative term in Eq. (10.1-22), depending on the coefficient values, can lead to a decrease in the phenol rate when a sufficient amplitude of oxygen and hydrogen partial pressures have been reached. Eq. (10.1-22) should thus be able to reproduce the observed experimental trend.

The experimental partial pressure dependence for both reactions could not be determined due to the impossibility to isolate the benzene hydroxylation reaction and phenol consumption to CO_2 in the system (the phenol rate measured at the reactor end is the result of both processes). A discussion, however, on the relative influence of the oxygen and hydrogen partial pressures on the phenol rate, depending on the value of the coefficient involves, can be carried out.

If the sum of the terms, including the partial pressures at the denominator, is much higher than 1, then the partial order for H_2 would be -0.5 in the phenol formation term and -7.5 for the phenol oxidation term. A higher negative exponent at the oxidation term would lead to a steeper decrease in the phenol rate (partial pressures have values <1, expressed in bar). The partial order for O_2, for the positive term of Eq. (10.1-24), would be -0.5, and -1.5 for the negative term. The same partial pressure dependence on H_2 and O_2 is found in the phenol formation term. A negative value of -0.5 leads, however, to a decrease in the phenol rate as the partial pressure increases; this does not model the increasing domain of the experimental profile.

If the sum of the terms, at the denominator, is much smaller than 1, then P_{O2} would have a partial order of 0.5 for the positive term and 6 for the negative term (less influential, lower ability to induce a decrease). P_{H2} would only be present in the phenol formation term at the exponent 0.5, and would also have a similar effect to P_{O2}.

Case B: R7 reversible

Eq. (10.1-24) applies where R7 is reversible. A negative component proportional to P_{H2} at the numerator contributes to a reduction in the amplitude of the phenol formation part of Eq. (10.1-24). The phenol formation term, however, remains at the same partial order for H_2 so that the influence of the partial pressures is the same as in case A. The reversibility of R7 might slightly affect the shape of the phenol curve but should not result in any significant change in the modeling of the phenol profile in comparison to case A.

<u>CO_2 rate:</u>

The proposed model defines the CO_2 formation rate as the sum of the benzene oxidation rate and the phenol oxidation rate, as indicated in Eq. (10.1-25). In both components of the CO_2 rate, oxygen partial pressure is present at the numerator at the exponent 6 and at the denominator at the exponent 7.5. Hydrogen is only present at the denominator at the exponent 7.5. Here again, the partial pressure dependence of the CO_2 rate is influenced by the value of the sum of the terms containing the partial pressures at the denominator.

If $\sqrt{K_{H2}P_{H2}} + \sqrt{K_{O2}P_{O2}} + K_4 \cdot \sqrt{K_{O2}P_{O2}} \cdot \sqrt{K_{H2}P_{H2}} + K_B P_B + K_{Ph}P_{Ph} + K_{H2O}P_{H2O} + K_{CO2}P_{CO2}$ is much higher than 1, then the partial order for oxygen would be -1.5 and the partial order for hydrogen would be -7.5. On the other hand, if this sum is small in comparison to 1, then O_2 would be at the partial order 6 and H_2 at the partial order null.

Experimentally, it was observed that the oxygen partial pressure influenced the CO_2 rate at a very low partial order of 0.31 (see Fig. 9.2-5), while the evolution of the CO_2 rate was, in fact, quasi independent of the hydrogen partial pressure (partial order null). With regard to the oxygen partial pressure, 0.31 is an intermediary value between -1.5 and 6, tending however, more towards the high values of the denominator of Eq. (10.1-25). When analyzing the influence of the hydrogen partial pressure, a partial order of 0 for hydrogen is obtained, as the sum of the denominator partial pressure terms would be negligible. This is in contradiction with what is observed with O_2.

Eq. (10.1-25) is able to describe only one of the 2 partial pressure trends, which were experimentally observed. For a modeled CO_2 profile matching the experimental CO_2 rate at the exponent 0.31 with varying oxygen partial pressures, the model will show a strong decrease in the CO_2 rate with increasing hydrogen partial pressures, instead of remaining stable.

It suggests itself that the model is not developed enough to reproduce the experimental results.

For a quantitative appraisal on the ability of the model to reproduce the observed experimental product rates, the unknown coefficients involved in Eq. (10.1-18), (10.1-20), (10.1-22), (10.1-24) and (10.1-25) should be accurately determined by running a regression on the experimental data, which minimizes the deviation between experimental and modeled data. However, as "only" 24 experimental measurements are available for the determination of 12 to 14 coefficients, the reliability of the calculation is limited. For this reason, no regression process was carried out on the experimental data; a screening method for coefficient determination was instead used.

Based on the literature data, some coefficients related to hydrogen oxidation to water over a Pd surface can be calculated and extrapolated for the present case, even though differences in terms of the used catalyst and temperature domain exist.

For simplification purposes, Eq. (10.1-18) and Eq. (10.1-22) were used as water and phenol rate models respectively in order to limit the number of unknown coefficients, which had to be determined. It was thus assumed that R6 and R7 were irreversible.

The remaining adsorption coefficients and kinetic constants were estimated via an undirected search on each value until a combination of coefficient satisfactorily fitted with the experimental data. An approximated kinetic could thus be implemented in a simple 1D model in Matlab, which supplied valuable information on the evolution of the product and educt concentration along the reactor axis. It was thus possible to simulate the evolution of the H_2/O_2 ratio in a double-membrane system, used for benzene hydroxylation, and draw a conclusion on the improved control of the reaction atmosphere thanks to the inclusion of a second membrane in the standard concept.

10.2.2. Estimation of adsorption constants and kinetic parameters for simulation purpose

Johansson et al. also investigated hydrogen oxidation with O_2 to water over a Pd surface using the Langmuir-Hinshelwood model; however, at much higher temperatures as they used microcalorimetry as a measurement technique (1027°C as opposed to the 150°C used here) [71]. In spite of the deviation in adsorption and desorption behaviors, related to the temperature difference and slightly different catalytic surface ($Pd_{90}Au_{10}$ wt.%), the expressions supplied by Johansson et al. could be employed for the determination of the adsorption constants by adjusting the temperature to that, which was applied in the experiments. The adsorption and desorption constants are defined as the following respectively:

$$k_{ads}=\frac{S}{\Gamma^m}\sqrt{\frac{k_BT}{2\pi M}} \quad \text{and} \quad k_{des}=A.\exp\left(-\frac{E_a}{k_BT}\right) \qquad \text{Eq. (10.1-23)}$$

with S the sticking coefficient, Γ the density of surface sites on which the gas molecule can stick (1.52×10^{19} m^{-2} for Pd), m the number of surface sites occupied by the adsorbed gas molecules, k_B the Boltzman's constant, T the temperature, M the molar mass of the gas molecule, A the pre-exponential coefficient and E_a the activation energy for desorption. Johansson et al. supply values for E_a, S and A (in the relevant units for k_{ads} and k_{des}) for several reactions [71], in particular for the R1, R2, R5, R6 and R8 reactions described in the previous chapter, which allow for a determination of K_{H2}, K_{O2}, K_{H2O} and k_{5+} as fixed parameters. The adsorption constants K were defined as the ratio between k_{ads} and k_{des} for the different adsorbed species. The 8 remaining parameters were approximated via value screening in order to obtain a match with the experimental data shown in Table 9.2-3 (no precise determination of these coefficients was carried out here). The coefficients involved in Eq. (10.1-18), (10.1-20) and (10.1-25), which were calculated using literature data, are shown in a yellow background in Table 10.1-1; the estimated coefficients appear in *italic script*.

Table 10.1-1: Calculated kinetic parameters using literature data (yellow) [71] and estimated parameters via undirected search (italic)

Adsorption Equilibrium constants (bar^{-1})		Kinetic constants (mol.h^{-1}.bar^{-n})	
K_{H2}	4.5	K_4	*100*
K_{O2}	1.3	$k7^+$	*5.20×10^{-4}*
K_B	5	$k5^+$	2.50×10^{-2}
K_{Ph}	5	$k6^+$	*7.30×10^{-3}*
K_{H2O}	5.3	$k10^+$	*$3.00 \times 10^{+15}$*
K_{CO2}	*10*	$k11^+$	*3.30×10^{-4}*

Fig. 10.1-1 provides an overview on the modeled water and phenol rates, which were obtained using 4 fixed parameters from among the 12 parameters. The fact that not all parameters have been determined leads to an imperfect matching between model and experiment. With the exception of the initial increase in the phenol rate, it was, however, observed that the model is able to reproduce the trend of the experimental data, as was put forward in the model discussion. The CO_2 rate model logically showed greater deviations from the experimental results in the case of the varying hydrogen partial pressure, as previously mentioned and for this reason not displayed

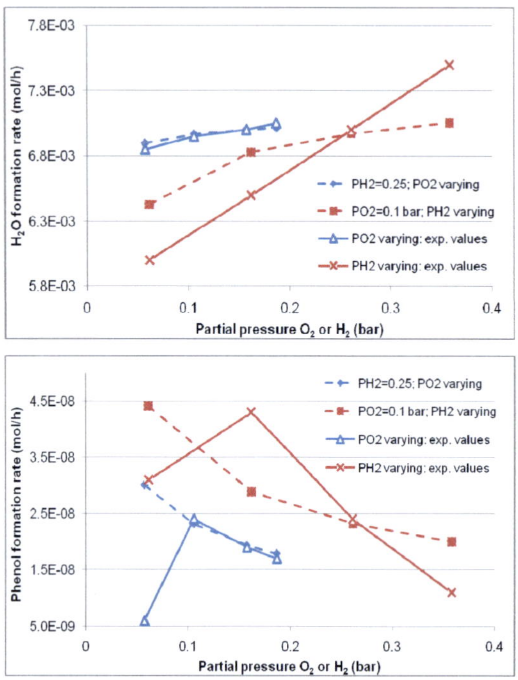

Fig. 10.1-1: Estimation of the missing parameters via a graphical comparison between the modeled and experimental data, here the water and phenol kinetics respectively. The imperfect matching of some modeled points with the measured data is shown. The modeled data appear as dashed lines.

The purpose of this was to allow for the implementation of reaction kinetic data in a 1-dimentional Matlab model of the reactor, which allowed for the simulation of the educt and product profiles inside the reactor (non-accessible information at the experimental level) and thus to better understand the working mode of the designed reactor.

11. Reactor simulation with Matlab 2007b

A model of the reactor system was realized in Matlab 2007b, which took into account the double membrane dosage of hydrogen and oxygen and the reactions observed in the system. The pseudo-homogeneity of the system and the plug flow behavior in the channels (same reaction development in all channels independently of their locations) are the basic assumptions for the produced 1-dimentional model. Benzene hydroxylation to phenol, benzene and phenol oxidation to CO_2 and direct water production were taken into account in the simulation. Their respective reaction kinetics based on Langmuir-Hinshelwood mechanisms, as modeled in part 10, were implemented in the model, using the coefficients given in Table (10.1-1). The real permeation behavior of the membranes, used for hydrogen and oxygen dosage, was also introduced into the model.

The evolution of the educt concentration (double-membrane dosage of hydrogen and oxygen) along the reactor axis can be observed together with the formation profile of the different products in the reactor. An interesting aspect of the simulation is that the "invisible" part of the experimental work, what happens along the reaction channels, can be simulated with Matlab.

Figs. 11-1-1 to 11.1-4 show the results of a simulation, which was carried out to reproduce experiments performed in double-membrane dosage with the following elements: PdAu-modified PdCu foil and UF-membrane installed in the reactor, reaction temperature of 150°C and H_2/O_2 ratio in the system close to 1. The obtained concentrations at the reactor exit correspond with the values observed in the experiments. The amounts of gas dosed into the reaction channels can be changed by modifying the permeation parameters available in the program (temperature and retentate pressure for O_2 and H_2).

Fig. 11.1-1 shows the simulated evolution of the H_2, O_2 and N_2 concentrations in the reactor.

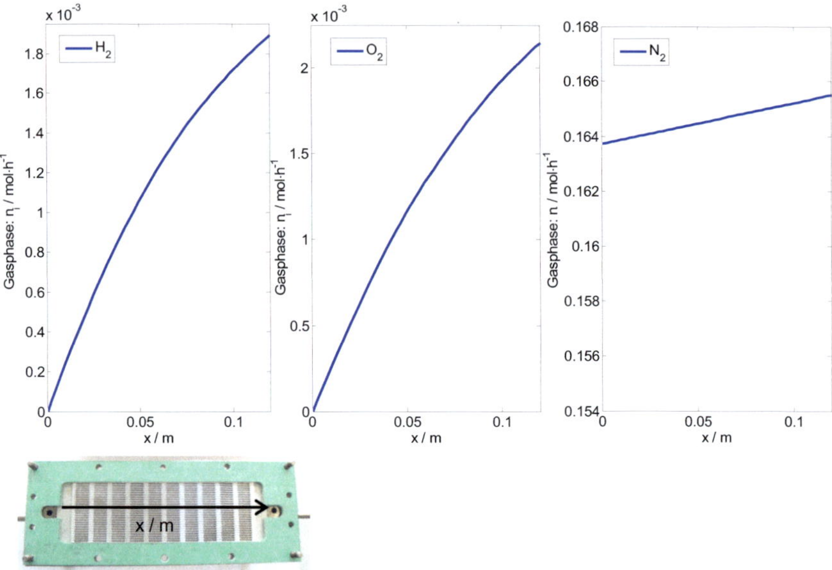

Fig. 11.1-1: Matlab simulation of gas molar flow along the reactor axis during reaction

Several observations can be made in this connection. Firstly, N_2 was used, in large amounts, as a carrier gas for benzene and was not involved in any reaction; however some nitrogen also diffused from the air flow passing through the porous UF membrane. A slight increase in the nitrogen concentration along the channel length is observed for this reason. A flat N_2 concentration profile would be obtained where a defect-free oxygen-selective membrane is simulated.

The hydrogen profile shows a regular increase in the concentration imposed by Sieverts's permeance all along the membrane length. The hydrogen concentration reaches its maximal value, of approx. 1.8×10^{-3} mol/h, at the reactor end as simulated here.

As expected, the oxygen concentration profile shows a similar behavior to hydrogen. It was introduced in the reaction area via a 2nd membrane acting over the whole reactor length; a porous UF-membrane and not a dense metal membrane was, however, used in this case. After a continuous concentration increase in the reactor, the gas phase exits the reactor with a molar oxygen rate of approx. 2.2×10^{-3} mol/h.

As a proof of concept of one of the key ideas in this project, Fig. 11.1-2 illustrates the evolution of the H_2/O_2 ratio along the reactor axis.

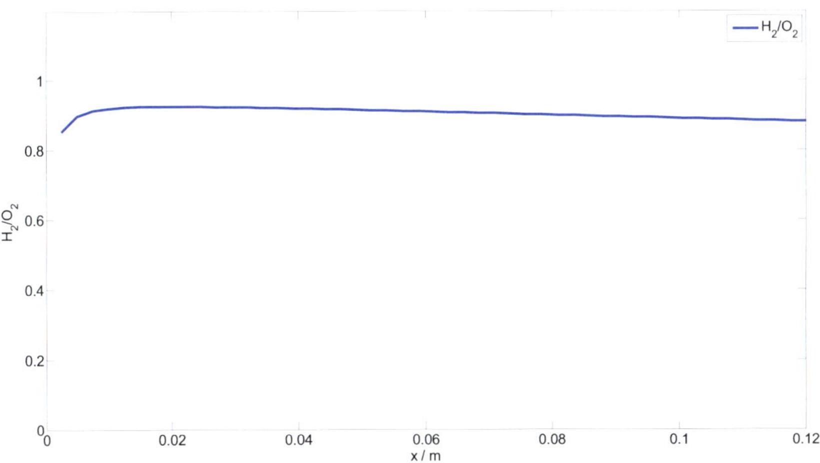

Fig. 11.1-2: Matlab simulation of the evolution of the H_2/O_2 concentration along the reactor axis during the process: proof of concept of the benefice of the double-membrane dosage on the control of the atmosphere composition during the reaction

One observes a quasi-constant H_2/O_2 ratio all along the reactor axis, with only a slight variation between values of 0.86 and 0.94. A stable ratio of both reaction gases was expected due to the introduction of a second membrane in the reactor in order to obtain a better control of the reaction atmosphere. This is in clear contrast to the membrane reactor developed by the AIST, which leads to opposite concentration profiles in the gas phase, creating 3 different reaction areas along the reactor axis.

The reactor simulation shows the benefit of the double-membrane dosage for achieving improved hydroxylation conditions in the reactor. This can be seen as a proof of concept of the double-membrane dosage.

Fig. 11.1-3 shows the benzene consumption along the reactor axis together with the phenol formation.

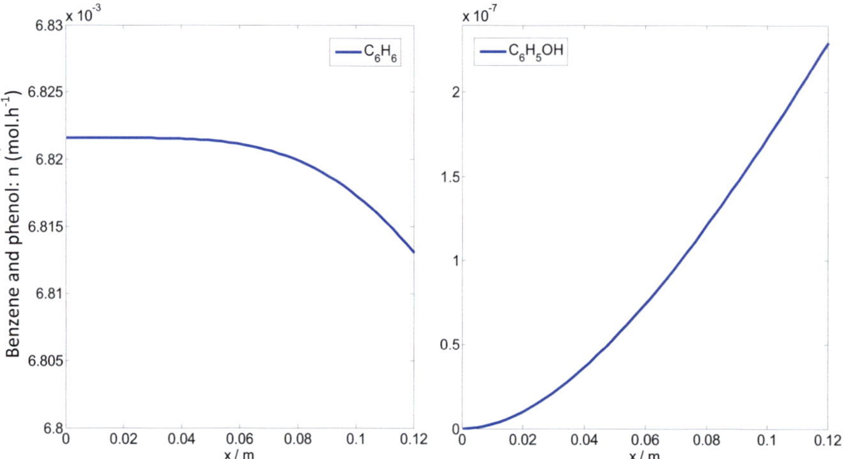

Fig. 11.1-3: Matlab simulation of benzene consumption and phenol formation along the reactor axis during the process

The low benzene consumption rate observed experimentally, which led to a very low conversion rate, is simulated here. The benzene concentration decreases from approx. 6.822×10^{-3} mol/h to 6.813×10^{-3} mol/h along the reactor. The benzene consumption appears to be more important in the second part of the reactor, where the atmosphere shows higher concentrations of hydrogen and oxygen (membrane dosage) available for the reaction.

The phenol profile shows a regular increase in the phenol formation along the channel, as soon as hydrogen and oxygen are introduced into the reactor; a concentration around 2.3×10^{-7} mol/h is reached at the exit.

Fig. 11.1-4 illustrates the evolution of the byproducts generated together with phenol: carbon dioxide and water

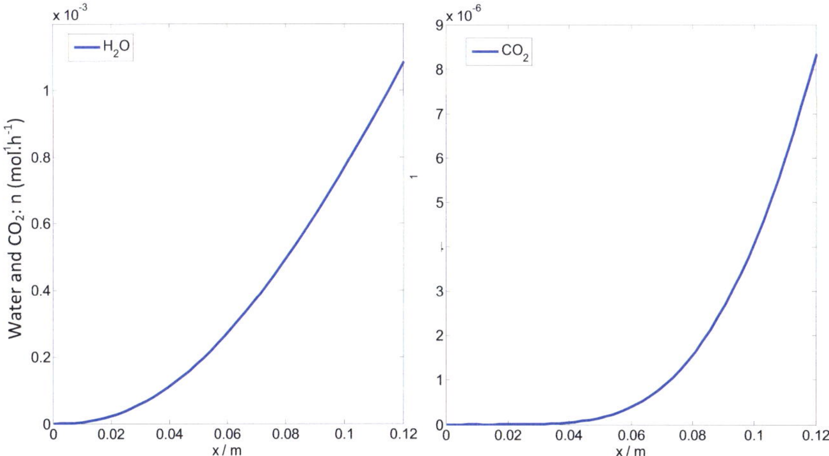

Fig. 11.1-4: Matlab simulation of byproduct formation (carbon dioxide and water) along the reactor axis during the process

As observed during the experiments, the water production rate (in the range of 1×10^{-3} mol/h) is much more important than the rate of any other product. It shows a regular increase all along the channel. The poor utilization of hydrogen and oxygen for phenol formation due to the large dominance of the water formation is apparent. The CO_2 concentration shows a nearly flat profile in the first 4 cm of the reactor (where the reactions to water and phenol could be privileged) before increasing significantly in the second part of the reactor, mainly due to benzene oxidation to CO_2, up to a molar ratio of 2.6×10^{-7} mol/h with the selected parameters. According to this simulation, the oxidation of benzene to CO_2 mainly takes place in the second part of the reactor.

The simulation section using Matlab, which implemented the experimentally measured permeation behaviors and reaction kinetics, underlines the benefit of double-membrane dosage for keeping the H_2/O_2 ratio constant in the channels, thus allowing for a better control of the atmosphere and for an enhanced hydroxylation reaction.

12. Conclusion

In this work, a double-membrane reactor was developed in order to perform and study the direct hydroxylation reaction of benzene to phenol in the gas phase, a direct reaction of high industrial interest, which was first presented by Niwa et al. in 2002 [24]. In this process, it is claimed that the surface of a compact Pd membrane is able to generate active hydroxyl species through the reaction between permeated hydrogen radicals and gas phase oxygen, which are, in turn, used to directly convert absorbed gas phase benzene into phenol. The innovation of the system lies in the use of the Pd membrane both for the supply of dissociated hydrogen atoms, under safe process conditions, and as a catalytic layer for the hydroxylation reaction itself. While very encouraging results were first reported by Niwa et al., i.e. a phenol yield exceeding 20% and a phenol selectivity above 77% attained at 200°C despite the formation of several byproducts (mentioned but not always quantified) [25], other researchers could not reproduce these results using a similar catalytic Pd membrane reactor, in particular due to the large amounts of water and CO_2 produced in the system [29]. In order to extend the hydroxylation area in the reaction, which was initially restricted by the competition between the oxidation and hydrogenation reactions due to opposite gas concentration profiles in a single-membrane system, we introduced a second membrane for the oxygen dosage which, combined with the H_2-selective Pd-based membrane allowed for a quasi-constant H_2/O_2 ratio along the reaction channel. In spite of the technical improvement, initial measurements, performed over a Pd-modified PdCu foil as the main membrane, led to a maximum C-based phenol selectivity of 9.6% and to a phenol rate of 1.3×10^{-8} mol/h at 150°C. Water formation due to the direct reaction between hydrogen and oxygen in system was also observed. A modification of the catalytic properties of the PdCu membrane surface was realized in order to increase the phenol selectivity of said membrane.

The results obtained in the second part of this project showed an improvement in the activity and selectivity of the PdCu membrane, which was used both as a catalyst and as a hydrogen distributor in a catalytic membrane reactor for the direct gas phase hydroxylation of benzene to phenol; this improvement was achieved by sputtering different catalytic layers onto the surface. The highest phenol productivity (1.5×10^{-4} $mol.h^{-1}.m^{-2}$) and carbon-based phenol selectivity (67%), among the three different systems studied, was reached at 150°C with a $Pd_{90}Au_{10}$ compact layer, which was 1 µm thick. By way of comparison, a 5 µm thick PdGa layer showed promising kinetic properties for the target reaction (restriction of the by-products rates, i.e. CO_2 and H_2O) but suffered from low hydrogen permeance, which limited the obtainable results. The production of thinner PdGa layers led to higher H_2 permeation rates and improved phenol selectivities but the achieved phenol production rates were still limited in comparison to those of $Pd_{90}Au_{10}$. The third system, namely discrete $Pd_{90}A_{10}$ particles on a discontinuous V_2O_5 layer, which was prepared by the reactive sputtering of vanadium followed by the sputtering of $Pd_{90}Au_{10}$ did not lead to the expected enhancement of the activity, which was based on the increase in the surface area. This assumption here is that no discrete particles were actually formed, rather a thin layer, which thus did not resulted in the desired increase in the surface area.

Even though there is certainly room for further improvements, for instance, by modifying the catalytic layers and optimizing the reactor design as well as the process conditions, the results at present do not support an industry application due to the poor utilization of the hydrogen and oxygen (main byproduct is water) as well as the low activity per unit volume (low catalyst activity and limited surface area per reactor volume). As had previously been demonstrated by other scientists [29; 63], the promising results initially reported by the Japanese group could not, in any case, be reproduced in the double-membrane reactor. Notwithstanding the disappointing performance limitation observed for phenol, the new reactor type used in this work did allow for a better understanding of the multiple gas phase reactions as they compete with each other in the system and the quantification of the reaction products. A reaction model based on gas phase adsorption and surface reaction mechanisms is also proposed in order to assess the product rates for a wide range of process parameters.

A final comparison can be made with a recent work carried out using a similar innovative system. In a recent paper by Ye et al. [64], a microsystem version of a Pd membrane reactor was proposed as the means to perform direct hydroxylation of benzene to phenol under optimized conditions. The concept behind this shares with this work the idea of shaping the reaction zone into an array of microchannels with the aim of improving the control of the reaction atmosphere throughout the reaction zone. Ye et al. used, however, silicon-based MEMS technology for the production of a microsystem, while a more standard mechanical (micro) structuring method was employed in this project. A distributed dosage of oxygen through a second membrane, in addition to the distributed supply of hydrogen through the Pd membrane, is proposed here [64] in order keep the H_2/O_2 ratio constant along the reactor length. The miniaturization of the system – the microchannels were 100 µm wide and 5 mm long – allows for a high surface to volume ratio, which is claimed to be a key advantage of the MEMS system as it allowed for a more efficient utilization of the active hydrogen species for the desired hydroxylation reaction [64]. High benzene conversion and phenol selectivity is reported, e.g. at 200°C, the phenol yield was 20% for a benzene conversion of 54% and a H_2/O_2 ratio of 2.5; this corresponds to a phenol selectivity of 37%. In contrast to this work, 2-hydroxycyclohexanone and p-benzoquinone were the main byproducts for H_2/O_2 ratios greater than 2, while CO_2 formation played only a minor role. Water formation is mentioned but once again not quantified.

Based on the results of this work and those of others [29; 61-63], the reaction atmosphere has tended to favor hydrogen oxidation to H_2O; at hydrogen partial pressures around 0.2 bar and a H_2/O_2 ratio, which varied from 1.5 to 5. The inefficiency of the benzene direct hydroxylation in connection with a Pd membrane was observed in these works. This is due to the predominance of water formation during the process, even though it was possible to positively influence the benzene reactivity in the used systems.

The analysis of the results obtained using the MEMS technology is simply too vague regarding the question, indeed large problem, of the direct reaction to water. For this reason, this system, similar to the Pd-membrane reactor solution (either using 1 or 2 membranes for the gas dosage), cannot be considered a serious and efficient alternative to the current industrial processes for phenol production.

13. References

[1] M. Iwamoto, J. Hirata, K. Matsukami, S. Kagawa, J. Phys. Chem. 87 (1983) 903.

[2] E. Suzuki, K. Nakashiro, Y. Ono, Chem. Lett. (1988) 953.

[3] G.I. Panov, V.I. Sobelev, A.S. Kharinov, J. Mol. Cat. 61 (1990) 85.

[4] J.S. Yoo, A.R. Sohail, S.S. Grimmer, C. Choi-Feng, Catal. Lett. (1994) 299.

[5] L.V. Pirutko, V.S. Chernyavsky, A.K. Uriarte, G.I. Panov, Appl. Catal. A 227 (2002) 271.

[6] B. Louis, L. Kiwi-Minsker, P. Reuse, A. Renken, Ind. Eng. Chem. Res. 40 (2001) 1454.

[7] R. Leanza, I. Rossetti, I. Mazzala, L. Forni, Appl. Catal. A 205 (2001) 93.

[8] A.Ribera, I.W.C.E. Arends, S. de Vries, J. Perez-Ramirez, R.A. Sheldon, J. Catal. 195 (2000) 287.

[9] Y.W. Chen, Y.H. Lu, Ind. Eng. Chem. Res. 38 (1999) 1893.

[10] M. Stockmann, F. Konietzni, J.U. Notheis, J. Voss, W. Keune, W.F. Maier, Appl. Catal. A 208 (2001) 343.

[11] D. Bianchi, R. Bortolo, R. Tassini, M. Ricci, R. Vignola, Angew. Chem. 39 (2000) 4321.

[12] S.K. Das, A. Kumar Jr., S. Nandrajog, A. Kumar, Tetrahedron Lett. 36 (1995) 7909.

[13] W. Zhang, J. Wang, T. tanev, T.J. Pinnavaia, Chem. Comm. (1996) 979.

[14] J. Nomiya, H. Yanagibayashi, C. Nozaki, K. Kondoh, E. Hiramatsu, Y. Shimizu, J. Mol. Catal. A: Chem. 114 (1996) 181.

[15] Y. Tang, J. Zhang, J. Serb. Chem. Soc. (2006) 111.

[16] T. Jintoku, H. Tanigushi, Y. Fujiwara, Chem. Lett. 193 (1987) 1865

[17] A. Kunai, T. Wani, Y. Uehara, F. Iwasaki, Y. Kuroda, S. Itoh, K. Sasaki, Bull. Chem. Soc. Jpn. 62 (1989) 2613.

[18] T. Tatsumi, K. Yuasa, H. Tominaga, Chem. Commun. (1992) 1446.

[19] H. Ehrich, H.Berndt, M. Pohl, K. Jahnisch, M. Baerns, Appl. Catal. A: Gen. 230 (2002) 271.

[20] T. Kitano, T. Wani, T. Ohnishi, J. Li-Fen, Y. Kuroda, A. Kunai, K. Sasaki, Catal. Lett. (1991) 11.

[21] J.S. Hwang, C.W. Lee, D.H. Ahn, H.S. Chai, S.E. Park, Res. Chem. Intermed. (2002) 527.

[22] T. Miyake, M. Hamada, H. Niwa, M. Nishizuka, J. Mol. Catal. A: Chem. 187 (2002) 199.

[23] W. Laufer, J.P.M. Niederer, W.F. Hoelderich, Adv. Synth, Catal. (2002) 344

[24] S. Niwa, M. Eswaramoorthy, J. Nair, N. Itoh, H. Shoji, T. Namba, F. Mizukami, Science 295 (2002) 105.

[25] N. Itoh, S. Niwa, F. Mizukami, T. Inoue, A. Igarashi, T. Namba, Catal. Comm. 4 (2003) 243.

[26] K. Sato, S. Niwa, T.a. Hanaoka, K. Komura, T. Namba, F. Mizukami, Catal. Letters 96 (2004) 107.

[27] K. Sato, T. Hanaoka, S. Niwa, C. Stefan, T. Namba, F. Mizukami, Catal. Today 104 (2005) 260.

[28] K. Sato, T. Hanaoka, S. Hamakawa, M. Nishioka, K. Kobayashi, T. Inoue, T. Namba, F. Mizukami, Catal. Today 118 (2006) 57.

[29] G.D. Vulpescu, M. Ruitenbeek, L.L. van Lieshout, L.A. Correia, D. Meyer, P.P.A.C. Pex, Catal. Comm. 5 (2004) 347.

[30] A.G. Knapton, Platinum Metals Rev. 21 (1977) 44.

[31] P.P. Mardilovitch, Y. She, Y.H. Ma, M.M. Rei, AIChE J. 44 (1998) 310

[32] US Patent Nr. 6,152,987, Ma et al., Nov.28, 2000

[33] Y.H. Ma, I.P. Mardilovitch, E.E. Engwall, Ann. N.Y. Acad. Sci. 984 (2003) 346.

[34] Y.H. Ma, B. C. Akis, M.E. Ayturk, F. Guazzone, E.E Engwall, I.P. Mardilovitch, Ind. Eng. Chem. Res. 43 (2004) 2936.

[35] C.G. Sonwane, J. Wilcox, Y.H. Ma, J. Chem. Phys. 125 (2006).

[36] P. Quicker, V. Höllein, R. Dittmeyer, Catal. Today 56 (2000) 21

[37] R. Dittmeyer, V. Höllein, K. Daub, J. Mol. Catal. A: Chem. 173 (2001) 135.

[38] V. Höllein, M. Thornton, P. Quicker, R. Dittmeyer, Catal. Today 67 (2001) 33.

[39] Y. Huang, R. Dittmeyer, J. Membr. Sci. 282 (2006) 296.

[40] J. Okazaki, D.A.P. Tanaka, M.A.L. Tanco, Y. Wakui, F. Mizukami, T.M. Suzuki, J. Membr. Sci. 282 (2006) 370.

[41] J.N. Keuler L. Lorenzen, R.D. Sanderson, V. Prozesky, W.J. Przybylowicz, Thin Solid Films 347 (1999) 91.

[42] S. Abate, G. Centi, S. Perathoner, F. Frusteri, Catal. Today 118 (2006) 189.

[43] H. Gao, J.Y.S. Lin, Y. Li, B. Zhang, J. Mem. Sci. 265 (2005) 142.

[44] X. Pan, M. Kilgus, A. Goldbach, Catal. Today 104 (2005) 225.

[45] L. Yuan, A. Goldbach, H. Xu, J. Phys. Chem. B (2007) 111.

[46] V.M. Gryaznov, S.G. Gul'yanova, S. Kanizius, Rus. J. Phys. Chem. 47 (1973) 1517.

[47] R.E. Coles, Brit. J. Appl. Phys. 14 (1963) 342.

[48] Diplomarbeit J. Bergwerff, University of Utrecht, „The interaction of O_2 with silver. A survey on the molecular processes in the passage of O_2 through a silver membrane" - available online

[49] P. Landon, P.J. Collier, A.F. Carley, D. Chadwick, A.J. Papworth, A. Burrow, C.J. Kiely, G.J. Hutchings, Phys. Chem. Chem. Phys. 5 (2003) 1917.

[50] J.K. Edwards. B. Solsona, P. Landon, A.F. Carley, A. Herzing, M. Watanabe, C.J. Kiely, G.J. Hutchings, J. Mater. Chem. 15 (2005) 4595.

[51] J.K. Edwards. B. Solsona, P. Landon, A.F. Carley, A. Herzing, C.J. Kiely, G.J. Hutchings, J. Catal. 236 (2005) 69.

[52] V. Schröder, B. Emonts, H. Janßen, H-P. Schulze, Chem. Ing. Technik 75 (2003) 914

[53] A. Goldbach, L. Yuan, H. Xu, Sep. Pur. Technol. (2010), doi:10.1016/j.seppur.2010.01.007.

[54] L. Yuan, A. Goldbach, H. Xu, J. Phys. Chem. B 112 (2008) 12692.

[55] L. Yuan, A. Goldbach, H. Xu, J. Membr. Sci. 322 (2008) 39.

[56] Fig. from "General experimental techniques" of PhD thesis "Design, manufacture and properties of Cr-Re alloys for application in satellite thrusters" (L. Gimeno-Fabra, ETSEIB Barcelona), taken from "Techniques de l'ingénieur"

[57] S. Benfer, P. Arki, G. Tomandl, Adv. Eng. Mat. 6 (2004) 495.

[58] S. Benfer, U. Popp, H. Richter, C. Siewert, G. Tomandl, Sep. Pur. Technol. 22-23 (2001) 231.

[59] U. Aust, S. Benfer, M. Dietze, A. Rost, G. Tomandl, J. Membr. Sci. 281 (2006) 463.

[60] G. Saracco, V. Specchia, Inorganic Membrane Reactors, chap. 17 of Structured Catalysts and Reactors (A. Cybulski, J. Moulijn), ISBN 0-8247-9921-6

[61] L. Bortolotto, R. Dittmeyer, Sep. Pur. Technol. 73 (2010) 51.

[62] M.H. Sayyar, R.J. Wakeman, Chem. Eng. Research and Design 86 (2008) 517.

[63] S. Shu, Y. Huang, X. Hu, Y. Fan, N. Xu, J. Phys. Chem. C 113 (2009) 19618.

[64] S.Y. Ye, S. Hamakawa, S. Tanaka, K. Sato, M. Esashi, F. Mizukami, Chem. Eng. J. 155 (2009) 829.

[65] K. Lemke, H. Ehrich, U. Lohse, H. Berndt, K, Jähnisch, Appl. Catal. A: General 243 (2003) 41.

[66] R.S.G. Ferreira, P.G.P. de Oliveira, F.B. Noronha, Applied Catal. B: Environmental 50 (2004) 243.

[67] G. Hamm, T. Schmidt, J. Breitbach, D. Franke, C. Becker, K. Wandelt, Surf. Sci. 562 (2004) 170.

[68] H. Orita, N. Itoh, Appl. Catal. A: General 258 (2004) 17.

[69] Y-G. Kim, J.E. Soto, X. Chen, Y-S. Park, M.P. Soriaga, J. Electroanalytical Chem. 554-555 (2003) 167.

[70] L.G. Petersson, H.M. Dannetun, I. Lundström, Surf. Sci.163 (1985) 273.

[71] A. Johansson, M. Försth, A. Rosen, Surf. Sci. 529 (2003) 247.

[72] G. Zheng, E.I. Altman, Surf. Sci. 462 (2000) 151.

[73] T. Engel, H. Kuipers, Surf. Sci. 90 (1979) 181.

[74] G. Pauer, M. Kratzer, A. Winkler, Vacuum 80 (2005) 81.

[75] PhD Thesis J. Osswald "Active-site isolation for the selective hydrogenation od acetylene: the Pd-Ga and Pd-Sn intermetallic compounds"

[76] K. Kovnir, M. Armbrüster, D. Teschner, T.V. Venkov, F.C. Jentoft, A. Knop-Gericke, Y. Grin, R. Schlögl, Sci. Technol. Adv. Mat. 8 (2007) 420.

[77] S.K. Gade, P.M. Thoen, J.D. Way, J. Membr. Sci. 316 (2008) 112.

[78] S.K. Gade, E.A. Payzant, H.J. Park, P.M. Thoen, J.D. Way, J. Membr. Sci. 340 (2009) 227.

[79] X. Zhang, G. Xiong, W, Yang, J. Membr. Sci. 314 (2008) 226.

[80] R. Dittmeyer, L. Bortolotto, Appl. Catal. A: Gen. (2010), doi:10.1016/j.apcata.2010.07.024

[81] N. N. Li, Advanced Membrane Technology and Applications, Wiley-Interscience, 2008, ISBN: 0471731676, chap. 34.3.1

14. Appendix

<u>14.1: Appendix 1: Programming of a YAG-Laser from *LASERPLUSS AG* for engraving V-shaped micro-channels in stainless steel plates</u>

The Inscript program contains different items for the setting of the parameters for engraving a plate, in our case the goal was to produce V-shaped gas microchannels. The main parameters are:

- the *laser focus* on the surface to be engraved,
- the laser *marking speed,*
- the activation of the *wobble function* (rotating movement of the laser beam), i.e. the selection of the frequency and radius of said,
- the *power level,*
- the *laser frequency* (Q-switch mode),
- the programming of the laser path to be followed (channel, square surface, etc.),
- the *repeat function*, which defines how many times the laser will follow the defined path,
- the filling mode in order to select the stripe structure (cross or single) made by the laser at the bottom of the engraved surface.

There are 6 main parameters, which influence the structure of the engraved material and which must be studied and optimized prior to the use of the laser. The goal is to produce gas microchannels of a specific shape and dimension, in this case V-shaped, with a surface width and depth of 200 µm. The purpose of these channels is to conduct the gas flow (H_2 and O_2) along the reactor under the membranes, which are, in turn, supported by the microstructured stainless plates.

Parameter study prior to the realization of the gas microchannels with the Nd:YAG laser

Different samples were produced on stainless steel segments by varying the laser parameters; the results were subsequently examined via a metallographic analysis of the structure of the lasered channels.

Figures 14.1 shows a top view of a lasered segment.

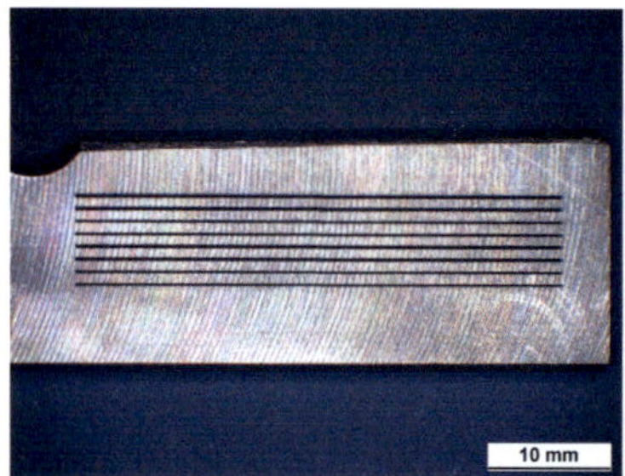

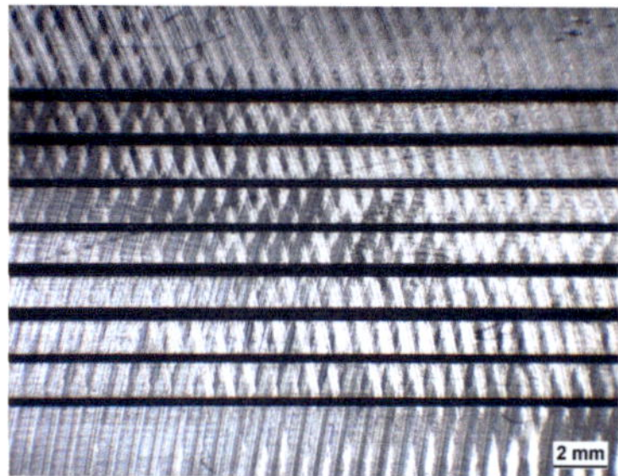

Fig. 14.1: View of a lasered stainless steel sample previous to metallographic analysis

More than 20 samples were realized and analyzed in order to determine the optimized parameter combination. The main parameter changes involved the laser power, frequency, marking speed, use of the wobble function and a repetition on the laser path in order to influence the depth of the channels. A special combination is particularly required for engraving V-shaped microchannels on stainless steel plates; many trials with the laser had, therefore, to be realized in order to determine the best parameters for the specific requirements (see Table 14-1).

Table 14-1: Presentation of the parameter changes and influence on the structure of lasered samples

sample n°	marking speed (mm/s)	laser power (%)	frequency (Hz)	wobble function	wobble freq. (Hz)	wobble radius (µm)	filling mode	channel width (µm)	repeat (x times)	shape & depth ok (yes/no)
1	100	80	4.000	off	500	30	cross	100/200	1	N
2	10*	80	4.000	off	500	30	cross	100/200	1	N
3	1	80	4.000	off	500	30	cross	100/200	1	N
4	100	80	4.000	off	500	30	cross	100/200	3	N
5	100	60	4.000	off	500	30	cross	100/200	3	N
6	100	60	8.000	off	500	30	cross	100/200	3	N
7	100	60	12.000	off	500	30	cross	100/200	3	N
8	100	60	24.000	off	500	30	cross	100/200	3	N
9	100	80	4.000	off	500	30	cross	100/200	3	N
10	100	80	4.000	off	500	30	cross	100/200	6	N
11	100	80	4.000	off	500	30	cross	100/200	10	N
12	100	80	4.000	on	500	30	cross	100/200	3	N
13	100	80	4.000	on	500	30	cross	100/200	6	N
14	100	80	4.000	on	500	30	cross	100/200	10	N
15	200	80	4.000	on	500	30	cross	100/200	3	N
16	100	80	4.000	on	500	30	cross	100/200	3	N
17	100	20	4.000	on	500	30	cross	100/200	15	N
18*	100	20	4.000	on	500	30	cross	100/200	30	Y
19	100	20	4.000	on	500	30	cross	100/200	50	Y
20	100	80	4.000	on	500	30	cross	200	3	N
21	100	60	4.000	on	500	30	cross	500	5	N
22	100	60	4.000	on	500	30	cross	300	4	N

*parameter change   *optimized parameters identified

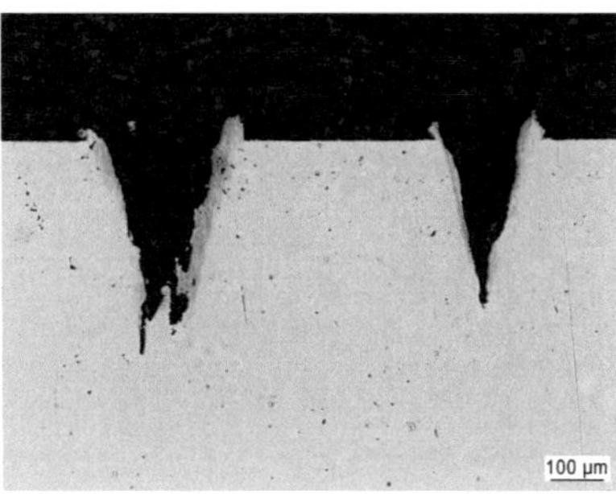

Fig 14-2: View of a sample engraved with optimized laser parameter: V-shape and desired dimensions have been produced

Two different channel dimensions have been produced on the sample in Fig. 16-2 (depth approx. 220 µm, surface width 200 and 100 µm respectively), both channels have the same geometry: V-shaped microchannels, which allowing for gas transport under a flat membrane have thus been produced.

14.2: Appendix 2: Matlab program for simulation of the double-membrane reactor for hydroxylation of benzene into phenol

```matlab
% Programm zur Simulation eines Membranreaktors für die
% Hydroxylierung von Benzol zu Phenol
% L. Bortolotto, 29.09.2009

% short g Format für Ausgaben im command window
format short g
% allgemeine Gaskonstante R, J/mol,K
R = 8.31441;

% Molanteile am Eintritt
% Komponentenreihenfolge
% Die Reaktionseite ist die Permeatseite
%       C6H6  C6H5OH  H2    O2    H2O   CO2   N2
y0 =  [0.04    0      0     0     0     0     0.96];

% Reaktorbedingungen
% Temperatur, K
T = 150 + 273.15;

% Druck Reaktorseite (=Permeatseite), Pa
pP = 1.0e+05;

% Druck Retentatseite H2, Pa
pRH2 = 3.0e+05;

% Druck Retentatseite O2, Pa
pRO2 = 3.0e+05;

% Volumenstrom am Eintritt, m3/h
vP0 = 100*1.0e-06*60;    % erster Wert = ml/min

% Stoffmengenströme am Eintritt, mol/h
n0(1:7) = y0(1:7)*pP*vP0/R/T;
```

```matlab
% Integrationsgebiet (Reaktorlänge) x, m
x0 = 0;      % Anfangskoordinate, m
xe = 0.12;   % Endkoordinate, m
nx = 50;     % Zahl der Stützstellen, -
xspan = linspace(x0,xe,nx);

% Lösung des DGL-Systems
options = odeset('RelTol',1.0e-08);
[x,n] = ode23s(@dgleichmix3,xspan,n0,options);
% Aufruf des Integrators

% Grafische Darstellung der Ergebnisse
figure(1)
subplot(1,2,1)
plot(x,n(:,1))
axis tight
xlabel('x / m')
ylabel('Festprodukte: n_i / mol\cdoth^{-1}')
legend('C_6H_6')

subplot(1,2,2)
plot(x,n(:,2))
axis tight
xlabel('x / m')
ylabel('Festprodukte: n_i / mol\cdoth^{-1}')
legend('C_6H_5OH')

figure(2)
subplot(2,2,1)
plot(x,n(:,5))
axis tight
xlabel('x / m')
ylabel('Festprodukte: n_i / mol\cdoth^{-1}')
```

```matlab
legend('H_2O')

subplot(2,2,2)
plot(x,n(:,6))
axis tight
xlabel('x / m')
ylabel('CO2: n_i / mol\cdoth^{-1}')
legend('CO_2')

figure(3)
subplot(3,3,1)
plot(x,n(:,3))
axis tight
xlabel('x / m')
ylabel('Gasphase: n_i / mol\cdoth^{-1}')
legend('H_2')

subplot(3,3,2)
plot(x,n(:,4))
axis tight
xlabel('x / m')
ylabel('Gasphase: n_i / mol\cdoth^{-1}')
legend('O_2')

subplot(3,3,3)
plot(x,n(:,7))
axis tight
xlabel('x / m')
ylabel('Gasphase: n_i / mol\cdoth^{-1}')
legend('N_2')

figure(4)
subplot(4,1,1)
plot(x,n(:,3)./n(:,4))
```

```matlab
axis tight
xlabel('x / m')
ylabel('H_2/O_2')
legend('H_2/O_2')

function f = dgleichmix3(x,n)
% Subroutine zur Berechnung der Differentialquotienten für
die Simulation
% eines Membranreaktors für die Hydroxylierung von Benzol
zu Phenol
% L. Bortolotto, 01.08.2010

% Querschnittsfläche A, m2
A = 4e-2*2e-03;

% Membranfläche pro Längeneinheit U, m
U = 4e-2;

% Druck Reaktionseite (=Permeatseite), Pa
pP = 1.0e+05;
% Druck Retentatseite H2 und O2, Pa
pR = 3.0e+05;

% Berechnung der Partialdrücke, Pa
ngP = sum(n(1:7));          % Gesamtstoffmengenstrom Permeat,
mol/h
p(1:7) = n(1:7)/ngP*pP; % Partialdrücke Permeat, Pa

% Adsorptionskonstanten
KH2 = 4.5/1.0e+5;
KO2 = 1.3/1.0e+5;
KB = 5/1.0e+5;
KPh = 5/1.0e+5;
KH2O = 5.3/1.0e+5;
```

```
KCO2 = 10/1.0e+5;
K4 = 100/1.0e+5;

% Berechnung von Thetaf
Thetaf = 1/(1 + (KH2*p(3))^0.5 + (KO2*p(4))^0.5 +
K4*((KH2*p(3))^0.5)*((KO2*p(4))^0.5) + KH2O*p(5) + KB*p(1)
+ KPh*p(2) + KCO2*p(6));

% Berechnung der anderen Theta
ThetaH = ((KH2*p(3))^0.5)*Thetaf;
ThetaO = ((KO2*p(4))^0.5)*Thetaf;
ThetaB = KB*p(1)*(Thetaf);
ThetaPh = KPh*p(2)*(Thetaf);
ThetaOH = K4*((KH2*p(3))^0.5)*((KO2*p(4))^0.5)*Thetaf;

% Geschwindigkeitskonstanten bei T=150°C, mol/Pan,h
k7 = 5.20e-4*1e5;
k11 = 3.30e-4*1e5;
k10 = 3.00e15*1e5;
k5 = 2.50e-2*1e5;
k6 = 7.30e-3*1e5;

% Hauptreaktion: C6H6 + H2 + O2 -> C6H5OH + H2O
r1 = k7*ThetaB*ThetaOH - k11*ThetaPh*(ThetaO)^12; %
Reaktionsgeschwindigkeit r1, mol/h

% Nebenreaktionen:

% NR1: 2H2 + O2 -> 2H2O
r2 = k5*ThetaOH*ThetaH + k6*(ThetaOH)^2; %
Reaktionsgeschwindigkeit r2, mol/h

% NR2: C6H6 + 15/2O2 -> 6CO2 + 3H2O
r3 = k10*ThetaB*(ThetaO)^12; % Reaktionsgeschwindigkeit
r3, mol/h
```

```
% NR3: C6H5OH + 7O2 -> 6CO2 + 3H2O
r4 = k11*ThetaPh*(ThetaO)^12; % Reaktionsgeschwindigkeit
r4, mol/h

% Stoffmengenänderungsgeschwindigkeit Ri, mol/h
Ri = [ -r1-r3          % C6H6
        r1-r4          % C6H5OH
       -r1-r2          % H2
       -r1-r2-r3-r4    % O2
        r2             % H2O
        r3+r4          % CO2
        0     ];       % N2

% Flussdichte ji, mol/m2,s
permH2 = 1.65e-07*1000; % Permeanz, mol/m2,s,Pa0,5
nH2 = 0.5;
jH2 = permH2*3600*0.002*((pR)^nH2 - p(3)^nH2); %
Flussdichte, mol/m,h
permO2 = 3.14e-07*10; % Permeanz, mol/m2,s,Pa
nO2 = 1;
jO2 = permO2*3600*0.002*(((pR/2)/5)^nO2 - ((p(4))/5)^nO2);
% Flussdichte, mol/m,h
permN2 = 3.14e-07; % Permeanz, mol/m2,s,Pa
nN2 = 1;
jN2 = permN2*3600*0.002*(((pR)*4/5)^nN2 -
((p(7))*4/5)^nN2); % Flussdichte, mol/m,h

ji = [ 0         % C6H6
       0         % C6H5OH
       +jH2      % H2
       +jO2      % O2
       0         % H2O
       0         % CO2
```

```
        +jN2 ];  % N2

f = Ri*A + ji*U;  % Differentialquotient dni/dx, mol/h,m
```

15. List of used symbols and abbreviations

Latin symbols

c_0	initial concentration of tracer	[mol.L^{-1}]
$c(t)$	tracer concentration	[mol.L^{-1}]
D	gas diffusion coefficient	[m^2.s^{-1}]
E, E_a	activation energy	[kJ.mol^{-1}]
E_i	height of discrete interval	[-]
$E(t)$	residence time distribution	[-]
F	gas flow in channel	[mL.s^{-1}]
$F(t)$	step function	[-]
h_{layer}	layer thickness	[m]
J	gas flux	[mol.m^{-2}.s^{-1}]
k^+	kinetic constant of direct reaction	[mol.h^{-1}.bar^{-n}]
k^-	kinetic constant of reverse reaction	[mol.h^{-1}.bar^{-n}]
k_{ads}	adsorption constant	[bar^{-1}]
k_B	Boltzman's constant: 1.38×10^{-23}	[J.K^{-1}]
k_{des}	desorption constant	[-]
K_i	adsorption equilibrium constant of educt i	[bar^{-1}]
m_{cat}	catalyst mass	[g]
M	molar mass	[g.mol^{-1}]
n	Sieverts's exponent	[-]
n_i, n_{i0}	mole flow and initial mole flow of educt i	[mol.h^{-1}]
N	number of intervals taken	[-]
$P\#$	intermediate interface pressure	[Pa]
P_{H2}, P_{O2}	partial pressure of hydrogen and oxygen	[bar]
P_{perm}	pressure at permeate side	[Pa]
P_{ret}	pressure at retentate side	[Pa]
P_{phenol}	phenol productivity	[mol.h^{-1}.g^{-1}]
ΔP	pressure difference	[Pa]
Q	membrane permeance	[mol.m^{-2}.s^{-1}.Pa^{-n}]

List of used symbols and abbreviations

Q'	membrane permeability	$[mol.m^{-1}.s^{-1}.Pa^{-n}]$
Q_0	pre-exponential factor	$[mol.m^{-2}.s^{-1}.Pa^{-n}]$
r_i, r_k	feed rate of educt i and rate of product k	$[mol.h^{-1}]$
R	universal gas constant: 8.314	$[J.mol^{-1}.K^{-1}]$
R_{Ph}, R_{CO2}, R_{H2O}	formation rates of phenol, CO_2 and H_2O	$[mol.h^{-1}]$
Re	Reynolds number	$[-]$
$S_{k,l}$	selectivity of product k based on educt i	$[\%]$
t	time	$[s]$
t_c	characteristic diffusion time across channel	$[s]$
Δt_i	width of discrete interval	$[s]$
$\bar{t}$	center of discrete interval	$[s]$
T	temperature	$[K]$
u, v, w	components of the fluid velocity field	$[m.s^{-1}]$
V	channel volume	$[mL]$
x_i	molar fraction of educt i	$[-]$
X_i	conversion of educt i	$[\%]$
Y_k	yield of product k	$[\%]$

Greek symbols

α	crystal phase of bcc structure	
β	crystal phase of fcc structure	
Γ	density of surface sites	$[m^{-2}]$
Θ_H	fraction of surface sites with adsorbed H species	$[-]$
Θ_f	fraction of free surface sites	$[-]$
λ	gas mean free path	$[nm]$
μ	dynamic viscosity	$[kg.m^{-1}.s^{-1}]$
μ_n	central moment of order n	$[s^{n+1}]$
ν'	kinematic viscosity	$[m^2.s^{-1}]$
ν_i, ν_k	stoichiometric coefficient of educt i	

	and product k	[-]
ρ	density	[kg.m^{-3}]
σ	variance around τ	[s]
τ	mean residence time	[s]

Abbreviations used

1D, 2D	1-dimensional, 2-dimensional
bcc	body centered cubic
DC	direct current
EDTA	ethylenediaminetetraacetic acid
EDX/EDAX	energy dispersion (analysis) of X-rays
ESMA	electron sample micro analysis
fcc	face centered cubic
FID	flame ionization detector
GC-MS	gas chromatograph - mass spectroscope
HEC	hydroxyethyl-cellulose
MCM	mesoporous crystalline material
MFC	mass flow controller
Nd:YAG	neodymium-doped yttrium aluminium garnet
NF	nano-filtration
PLOT	porous layer open tubular
PTV	programmed temperature vaporizer
RF	radio frequency
RTD	residence time distribution
SEM	scanning electron microscope
TCD	thermal conductivity detector
TEM	transmission electron microscope
TS	titanosilicate
UF	ultra-filtration

WCOT	wall coated open tubular columns
XRD	X-ray diffraction
ZSM	zeolite sieve of molecular porosity
µGC	micro gas chromatograph